Conversion to On-Site Sodium Hypochlorite Generation

Water and Wastewater Applications

Conversion to On-Site Sodium Hypochlorite Generation

Water and Wastewater Applications

Leonard W. Casson • James W. Bess, Jr.

LEWIS PUBLISHERS

A CRC Press Company
Boca Raton   London   New York   Washington, D.C.

Library of Congress Cataloging-in-Publication Data

Casson, Leonard W.
    Conversion to on-site sodium hypochlorite generation : water and wastewater applications / Leonard W. Casson, James W. Bess, Jr.
      p. cm.
    Includes bibliographical references and index.
    ISBN 1-58716-094-3 (alk. paper)
    1. Hypochlorinators. 2. Water—Purification—Disinfection. 3. Sewage—Purification—Disinfection. I. Bess, James W. II. Title.

TD463 .C37 2002
628.1′662—dc21   2002190726

This book contains information obtained from authentic and highly regarded sources. Reprinted material is quoted with permission, and sources are indicated. A wide variety of references are listed. Reasonable efforts have been made to publish reliable data and information, but the author and the publisher cannot assume responsibility for the validity of all materials or for the consequences of their use.

Neither this book nor any part may be reproduced or transmitted in any form or by any means, electronic or mechanical, including photocopying, microfilming, and recording, or by any information storage or retrieval system, without prior permission in writing from the publisher.

The consent of CRC Press LLC does not extend to copying for general distribution, for promotion, for creating new works, or for resale. Specific permission must be obtained in writing from CRC Press LLC for such copying.

Direct all inquiries to CRC Press LLC, 2000 N.W. Corporate Blvd., Boca Raton, Florida 33431.

Trademark Notice: Product or corporate names may be trademarks or registered trademarks, and are used only for identification and explanation, without intent to infringe.

Visit the CRC Press Web site at www.crcpress.com

© 2003 by CRC Press LLC

No claim to original U.S. Government works
International Standard Book Number 1-58716-094-3
Library of Congress Card Number 2002190726
Printed in the United States of America 1 2 3 4 5 6 7 8 9 0
Printed on acid-free paper

Dedication

To Jenny for her patience and support during my first book-writing endeavor.

Jim

Preface

This book addresses the design and operation of on-site sodium hypochlorite generation systems and their application for disinfection in water and wastewater treatment facilities. Prior to discussing the details of these systems, we review the origins of sodium hypochlorite generation. The current and pending regulations governing the use of chlorine are delineated in Chapter 2. Chapter 3 provides a brief overview of disinfection alternatives as applied to water and wastewater treatment. Chapter 4 explains the chemistry of disinfection. Chapter 5 provides an in-depth discussion of electrolyzer systems and the components of these systems. In Chapter 6 we highlight issues and problems to approach with caution when designing electrochlorination systems. Chapter 7 provides the reader with economic evaluation principles for electrolysis systems. Chapter 8 is intended to provide the reader with practical electrochlorination system installation, operation, and maintenance requirements. We examine system design and trouble analysis in Chapter 9. Finally, Chapter 10 outlines system safety concerns associated with on-site sodium hypochlorite generation systems.

Acknowledgments

I appreciate the support and understanding of Susan, Andrew, and Emily Jo in this writing endeavor. I am grateful for the advice of my colleagues and the prayers of my family, which helped bring this project to completion.

Leonard W. Casson

About the authors

Leonard W. Casson, Ph.D., P.E., D.E.E., is an associate professor of environmental engineering in the department of civil and environmental engineering at the University of Pittsburgh. Dr. Casson's research focus is the adsorption, fate, and transport of particles, chemicals, and environmental pathogens in unit processes and the natural environment. Dr. Casson is a professional engineer in Florida and Pennsylvania. He has also received specialty certification as a diplomate in the specialty of water supply and wastewater by the American Academy of Environmental Engineers. Dr. Casson is the author of over 50 papers in the water and wastewater treatment area.

James W. Bess, Jr., is a director of technology at Electrolytic Technologies Corporation in Aventura, FL. He has worked worldwide in the fields of dimensionally stable anodes, separated and unseparated electrolyzer cell research and development, and on-site sodium hypochlorite system engineering and installation for over 30 years. He also holds several electrolyzer cell-related patents.

Table of contents

Chapter 1 Introduction 1
 1.1 Origins of sodium hypochlorite generation..................................1
 1.2 Bleach...2
 1.3 On-site generation of sodium hypochlorite 2
 1.4 Dimensionally stable anodes..2
   References 3

Chapter 2 Federal regulations and pending regulations 5
 2.1 USEPA clean air act risk management plan.................................5
 2.2 USEPA office of pesticide programs..6
 2.3 OSHA process safety management standard 7
 2.4 Local and state regulations 8

Chapter 3 Disinfection applications and alternatives 9
 3.1 Chlorine gas..9
 3.2 Bulk manufactured sodium hypochlorite 10
 3.3 Ozone..13
 3.4 Chlorine dioxide 13
   References 13

Chapter 4 Disinfection chemistry 15
 4.1 Chlorine application chemistry...15
   4.1.1 Basic principles of disinfection 15
   4.1.2 Chemistry of elemental chlorine
          and sodium hypochlorite in water..................................15
   4.1.3 Chlorine demand reactions..17
     4.1.3.1 Demand reactions with ammonia......................18
     4.1.3.2 Demand reactions with organic
              and inorganic matter 19
   4.1.4 Disinfection kinetics..19
 4.2 On-site sodium hypochlorite generation chemistry 20
   4.2.1 Electrolytic cell reactions..21
   References 22

Chapter 5		Electrolyzer systems		25
5.1		Electrolyzer system types and principles of operation		25
5.2		Brine system: general description		25
5.3		Seawater system: general description		27
5.4		Electrolytic cells		28
	5.4.1	Tubular and plate electrolytic cell designs		29
		5.4.1.1	ELCAT — Chloromat	30
		5.4.1.2	PEPCON — ChlorMaster™	31
		5.4.1.3	Diamond Shamrock — Sanilec®	32
		5.4.1.4	DeNora — SEACELL®	33
		5.4.1.5	Mitsubishi Heavy Industries — Marine Growth Preventing System	33
		5.4.1.6	Daiki Engineering — Hychlorinator	34
	5.4.2	Cell module groupings		34
		5.4.2.1	Vertical cell circuits — seawater electrolysis	34
		5.4.2.2	Horizontal cell circuits — seawater electrolysis	35
		5.4.2.3	Horizontal cell circuits — brine electrolysis	36
5.5		Electrolysis systems		37
	5.5.1	Equipment		37
	5.5.2	Instruments		38
5.6		Materials of construction		39
	5.6.1	Plastic materials		39
		5.6.1.1	PVC (polyvinyl chloride) and CPVC (chlorinated polyvinyl chloride)	39
		5.6.1.2	Polypropylene (PP)	40
		5.6.1.3	Acrylonitrile butadiene styrene (ABS)	41
		5.6.1.4	Polyvinylidene fluoride (PVDF, Kynar®) and polytetrafluoroethylene (Teflon®)	41
		5.6.1.5	FRP (fiberglass reinforced plastic)	42
	5.6.2	Elastomeric sealing materials		42
		5.6.2.1	Fluorocarbon elastomer (Viton®)	42
		5.6.2.2	Ethylene propylene diene methylene (EPDM)	42
		5.6.2.3	Buna-N	42
		5.6.2.4	Neoprene	43
	5.6.3	Metals		43
		5.6.3.1	Titanium (Ti)	43
		5.6.3.2	Hastelloy® C 276 (Hast. C)	43
		5.6.3.3	316L stainless steel (316L)	44
5.7		DC power rectifiers		44
	5.7.1	Rectifier metering, control, and operation		45
		5.7.1.1	Cooling	45
		5.7.1.2	Area classification	47
		5.7.1.3	Maintenance	47
	5.7.2	DC rectifier operating status		48
		5.7.2.1	DC rectifier independent alarm conditions	48

	5.7.3	DC rectifier safety equipment	48
5.8	Control panel	49
	5.8.1	Instrumentation	49
5.9	Pressure and differential pressure equipment	50
5.10	Liquid flow equipment	51
5.11	Cell level and temperature	52
5.12	Water and brine instrumentation	53
5.13	Inlet seawater strainers	53
5.14	Brine system water softening	55
5.15	Temperature sensing equipment	57
5.16	Cell level and temperature	58
5.17	Tank level equipment	58
5.18	Salt storage-dissolver tanks	60
	5.18.1	Dissolver level controls	60
	5.18.2	Internal distribution and brine removal piping	61
	5.18.3	Salt addition	61
5.19	Product storage	62
	5.19.1	Tanks	62
5.20	Pump equipment	63
	5.20.1	Seawater booster pumps and seawater hypochlorite dosing pumps	63
	5.20.2	Brine system product dosing	65
5.21	Pipe, valves, and fittings for electrolysis systems	67
	5.21.1	PVC (polyvinyl chloride) and CPVC (chlorinated polyvinyl chloride)	67
	5.21.2	Polypropylene	67
	5.21.3	Acrylonitrile butadiene styrene (ABS)	67
	5.21.4	Polytetrafluoroethylene (Teflon®)	67
	5.21.5	FRP (fiberglass reinforced plastic)	67
5.22	Hydrogen handling practices	68
5.23	Brine system applications	70
	5.23.1	Cooling system applications	70
	5.23.2	Food processing	71
	5.23.3	Beverage operations	71
	5.23.4	Water disinfection	71
	5.23.5	Swimming pools	72
	5.23.6	Cyanide destruction	72
	5.23.7	Industrial bleaching	72
	5.23.8	Odor control	72
5.24	Seawater system applications	73
	5.24.1	Power stations	73
	5.24.2	Oil field water	73
	5.24.3	Cooling system applications	73
	5.24.4	Oil platforms	73
	5.24.5	Mammal pools	73

Chapter 6 Electrolysis system design considerations 75
    6.1   Brine hypochlorite systems...76
    6.2   Seawater hypochlorite systems..77
            6.2.1   Local regulations and codes 78
            6.2.2   Site type...78
            6.2.3   Local environmental conditions.............................78
            6.2.4   Available water pressure..79
            6.2.5   Dosing point pressure requirements......................79
            6.2.6   Dosage control methods...80
            6.2.7   Power supply requirements 80
            6.2.8   System sizing..80
            6.2.9   Biofouling control...82

Chapter 7 Economic evaluation principles for electrolysis systems 83
    7.1   System installation 83
    7.2   System operations...84
            7.2.1   Operating economics 84
                    7.2.1.1   Brine system operating costs 84
                    7.2.1.2   Seawater system operating costs 86

Chapter 8 Electrolysis system installation, operation, and maintenance..87
    8.1   Brine systems: general commissioning procedure......................91
    8.2   Seawater systems: general commissioning procedure...............92

Chapter 9 System design and trouble analysis 95

Chapter 10 System safety 99
    10.1   Chemical safety..99
            10.1.1   Sodium hypochlorite handling (NaOCl)..........99
            10.1.2   Hydrochloric acid handling (HCl) 100
    10.2   Electrical safety 101
    10.3   First aid 101
            10.3.1   Eye Burns — acid and alkali materials 101
            10.3.2   Skin burns — sodium hypochlorite, acid, or alkali materials...102
            10.3.3   Ingestion or gassing — sodium hypochlorite or alkali materials...102
            10.3.4   Electrical shock 102

Technical Standard Documents 105

Index 199

chapter 1

Introduction

1.1 Origins of sodium hypochlorite generation

Chlorine gas was first prepared in 1774 by Karle Scheele of Sweden. It was not until 1810, however, that Humphry Davy declared it an element before the Royal Society of London (White, 1999). Davy proposed the name of chlorine based on the Greek word *chloros* translated to green, greenish yellow, or yellowish green. The gas was liquified by compression in 1805 by Thomas Northmore (White, 1999).

In 1883, Faraday postulated the laws governing the passing of electric current through an aqueous salt solution, coining the word electrolysis (White, 1999). These fundamental laws are:

- The weight of a given element liberated at an electrode is directly proportional to the quantity of electricity passed through the solution. The electrical unit quantity is the coulomb.
- The weights of different elements liberated by the same quantity of electricity are proportional to the equivalent weights of the elements.

The first commercial production of chlorine began in 1890 by the Elektron Company (now Fabwerke-Hoechst A.G.) of Griesheim, Germany (White, 1999). The first electrolytic plant in the United States was started at Rumford Falls, Maine in 1892. In 1894, the Mathieson Chemical Company acquired the rights for the Castner mercury cell for the manufacture of bleaching powder at a demonstration plant in Saltville, Virginia (White, 1999). This facility subsequently moved to Niagara Falls, New York in 1897 and operated successfully until 1960 (White, 1999).

At first, the original electrolytic process was used primarily for caustic production. In 1909, the first commercial manufacturing of liquid chlorine began. This liquid was stored in 100-pound cylinders supplied from Germany. Tank cars were manufactured in the United States in 1909, as well, with a capacity of 15 tons. It was not until 1917 that 1-ton containers came into use for the U. S. military (White, 1999).

1.2 Bleach

Bleach was used as a bleaching agent starting in 1785 by mixing Scheele's gas in water and adding caustic potash (White, 1999). In 1789, Berthollet Tennent produced another liquid bleaching agent termed chlorinated milk of lime. This original product was greatly improved when it was dried to form bleaching powder (White, 1999).

Commercial bleach manufacture for domestic use was begun by the Mathieson Chemical Company in 1897 as a by-product of caustic manufacture to utilize some of the excess chlorine from the process. Chlorine that was not used in the bleach process or to manufacture hydrochloric acid was dumped into the Niagara River as a waste material (White, 1999).

1.3 On-site generation of sodium hypochlorite

While on-site generation was feasible at the turn of the 20th century, the limitation was the electrode materials, carbon or platinum, which would dissolve in service causing cell damage and poor product quality or were too expensive for practical use. No cell developed during this period provided reliable on-site sodium hypochlorite generation. Not until the development of the dimensionally stable anode for the chlorine industry in 1967 by an independent Belgian scientist, Henry Beer, was a reliable economic on-site generation cell practical (White, 1999). In 1971 J.E. Bennett, using the dimensionally stable anode, developed an unseparated electrolytic cell that was patented by Diamond Shamrock Corporation. Many variations in the electrodes and cell configurations have become available in the marketplace during the ensuing 25 years for electrolysis of both dissolved salt solution and seawater as the system feed stock.

1.4 Dimensionally stable anodes

The primary difference in the production of chlorine gas and on-site generation of bleach is the separation of the two basic cell products. In the production of chlorine the gas is separated on the anode side of the cell from the caustic (sodium hydroxide) solution produced on the cathode side through the reaction of sodium with feed water releasing hydrogen as a by-product gas. Each product is packaged separately when removed from the generation cell for use in unrelated processes such as pulp and paper bleaching using chlorine and caustic, chlorinating hydrocarbons for products such as PVC, and caustic used for glass manufacture. On-site generating cells do not separate the chlorine and caustic; thus the two products react together to form weak bleach solution up to 1.2% strength, also releasing by-product hydrogen gas.

References

White, G.C., (1999) *Handbook of Chlorination and Alternative Disinfectants,* Fourth Edition, Wiley Interscience, New York.

chapter 2

Federal regulations and pending regulations

The implementation of the Clean Air Act (CAA) risk management plan (RMP) for the storage of hazardous chemicals by the U.S. Environmental Protection Agency (USEPA) (June 1999) and the reregistration of chlorine gas by USEPA office of pesticide programs as a pesticide (Spring 2001) have accelerated the use of liquid sodium hypochlorite in the water and wastewater treatment industry. Below is a brief discussion of each of these regulations.

2.1 USEPA clean air act risk management plan

Congress passed the Clean Air Act Amendments of 1990. One section of these regulations required the USEPA to publish regulations and guidance for chemical accident prevention at facilities using extremely hazardous substances. The risk management program rule (RMP rule) implemented Section 112(r) of the CAA amendments. The RMP rule was written based upon existing industry codes and standards. This rule requires companies of all sizes that use certain flammable and toxic substances to develop a risk management program, which includes all of the following:

- A hazard assessment that details the potential effects of an accidental release, an accident history of the last five years, and an evaluation of worst-case and alternative accidental releases.
- A prevention program that includes safety precautions and maintenance, monitoring, and employee training measures.
- An emergency response program that spells out emergency health care, employee training measures, and procedures for informing the public and response agencies (e.g., the fire department) should an accident occur.
- A summary of the facility's risk management program must be submitted to the USEPA by June 21, 1999, which will make the

information available to the public. The plans must be revised and resubmitted every five years.

The risk management program was created to reduce chemical risk at the local level. The information included in the plan was intended to help local fire, police, and emergency response personnel (who must prepare for and respond to chemical accidents), and be useful to citizens in understanding the chemical hazards in communities. The USEPA anticipates that making the RMPs available to the public will stimulate communication between industry and the public to improve accident prevention and emergency response practices at the local level. A complete description of the risk management plan can be found on the World Wide Web at http://www.epa.gov/swercepp/bi-rima.htm.

The RMP defines threshold quantities (in pounds) of chemicals stored at a site as shown in Table 2.1. Exceeding these quantities at a single location requires the development of a risk management plan. A complete listing of chemicals and threshold quantities can be found at http://www.epa.gov/ceppo/caalist.html.

Table 2.1 Threshold Quantities for the Clean Air Act Risk Management Plan

Chemical	Threshold Quantity (Pounds)
Chlorine	2,500
Chlorine dioxide	1,000
Chlorine monoxide	10,000
Chlorine oxide	10,000

2.2 USEPA office of pesticide programs

On February 22, 1999, the USEPA office of pesticide programs (OPP) http://www.epa.gov/pesticides/ issued a reregistration eligibility decision (RED) on the pesticide chlorine gas. The RED required chlorine gas registrants to submit a revised labeling plan to the USEPA. This RED also allowed for re-examination of the required training and education procedures for people handling and dispensing chlorine gas. Although completion of this process was anticipated early in 2001, it was still incomplete as of October 2002.

The USEPA OPP has received and examined comments by several trade groups, state governments, universities, and individuals. The antimicrobials division of the OPP is still in the process of obtaining information from state agencies concerning possible methods of chlorine gas handling and dispensing training program implementation. Once the information gathering process is completed, a final decision on chlorine gas reregistration is expected.

chapter 2: Federal regulations and pending regulations

The current status of the reregistration decision may be found using the URL http://www.epa.gov/pesticides/reregistration/status.htm

The outcome of this RED impacts the water and wastewater treatment industry as follows:

- To promote uniformity, the USEPA is re-examining the training required for all people handling and dispensing chlorine gas. Therefore, the training for water and wastewater treatment plant operators using chlorine gas may be affected or altered.
- Although the USEPA OPP would prefer not to require additional training (above that presently required) for water and wastewater operators, each state must determine how the RED will be implemented.
- No action regarding the changing training requirements may be completed until the USEPA OPP finalizes the reregistration process.

In addition to the training issues discussed above, possible public concern over the treating of drinking water with a chemical classified as a pesticide may be substantial. This increased concern may be the result of a perceived increased risk.

2.3 OSHA process safety management standard

The Occupational Safety and Health Administration (OSHA) process safety management standard (PSM) of highly hazardous chemicals; explosives and blasting agents (29 CFR 1910.119) is a set of procedures in thirteen management areas designed to protect worker health and safety in case of accidental chemical releases. This rule is similar to the USEPA RMP. See http://www.osha-slc.gov/SLTC/processsafetymanagement/

Chlorine gas is regulated by both the USEPA RMP (discussed earlier) and OSHA PSM. On-site generated sodium hypochlorite is not included in either regulation. The threshold limit for chlorine gas to be regulated by OSHA PSM is 1500 pounds. This threshold limit (shown in Table 2.2) is 1000 pounds less than the 2500-pound threshold limit specified in the USEPA risk management plan.

Table 2.2 Threshold Quantities for the OSHA Process Safety Management Standard

Chemical	Threshold Quantity (Pounds)
Chlorine	1500
Chlorine dioxide	1000
Chlorine pentafluoride	1000
Chlorine trifluoride	1000

2.4 Local and state regulations

In addition to the above federal regulations, state regulators and local enforcement agencies may have more stringent regulations for the storage and transport of hazardous chemicals. As an example, the Los Angeles County Fire Department, which enforces the hazardous materials state reporting requirements, developed for the state of California and applied within Los Angeles County states that "a mixture that contains one percent or more of a hazardous ingredient is a hazardous material."

chapter 3

Disinfection applications and alternatives

Many chemical alternatives exist for the disinfection of water and wastewater treatment facilities. This publication addresses on-site sodium hypochlorite generation, however, this chapter will briefly address some of the other current chemical disinfection alternatives and their applications. Although ultraviolet light is becoming a more common disinfectant for wastewater and some water treatment facilities, it will not be included in the following discussion of chemical disinfectants.

3.1 Chlorine gas

Previously, water and wastewater treatment disinfection in the United States was almost exclusively associated with chlorine gas. Many events have contributed to a move away from the application of gaseous chlorine in water and wastewater treatment plants. First and foremost was the link between gaseous chlorine application and the formation of trihalomethanes (THMs) in the potable water distribution systems or wastewater outfalls. Other chemicals (e.g., ozone or chlorine dioxide) have been used to supplement or replace gaseous chlorine in some applications. However, the inexpensive nature of this disinfectant and its ease of application allowed it to be a primary disinfectant in the United States.

The implementation of the Clean Air Act (CAA) risk management plan (RMP) for the storage of hazardous chemicals by the USEPA (June 1999) and the reregistration of chlorine gas by USEPA office of pesticide programs as a pesticide (Spring 2001) have accelerated the use of liquid sodium hypochlorite in the water and wastewater treatment industry. Some utilities were reluctant to become involved with the development, implementation, and updating of a risk management plan required for continued use of chlorine gas as a disinfectant.

3.2 Bulk manufactured sodium hypochlorite

The majority of utilities in the United States transitioned from using gaseous chlorine to the bulk manufactured sodium hypochlorite. Commercial grade sodium hypochlorite can be produced by chemical manufacturers at concentrations as high as 16 weight % chlorine. However, typical commercial grade sodium hypochlorite concentrations are between 12 and 15 weight % chlorine. Concerns associated with the use of commercially produced sodium hypochlorite include high transportation costs and a lack of long-term stability. In addition, safety concerns exist and are associated with the handling, dosing, dilution, and accidental spilling or release of commercial grade sodium hypochlorite.

The stability of a sodium hypochlorite solution is dependent upon the following:

- The concentration of the hypochlorite solution
- The alkalinity and pH of the hypochlorite solution
- Storage temperature of the hypochlorite solution
- Concentration of impurities (e.g., Ni^{2+} and Cu^{2+}) that may catalyze hypochlorite decomposition
- Exposure to sunlight

The rate of reduction in the strength of a sodium hypochlorite solution increases with increasing hypochlorite strength, increasing solution temperature, and increasing holding time. The half-life of sodium hypochlorite (time to reach half of the original concentration) at ambient temperatures varies between 60 and 1700 days for solutions of 18 and 3%, respectively (Baker, 1969; Laubusch, 1963). Based on these data, a factor of six reduction in sodium hypochlorite concentration (18 to 3%) resulted in an almost 30-fold increase in sodium hypochlorite stability.

Gordon et al. (1997) also presented data on sodium hypochlorite stability and chlorate ion formation as shown in Table 3.1. In this study, commercially produced sodium hypochlorite was used at the original delivered strength and diluted to 10 and 5% solutions. These solutions were held in glass containers in the dark at 13 and 27 °C for 28 days.

The above data indicates that, as expected, the free available chlorine concentration reduction increases with increasing sodium hypochlorite strength and increasing temperature. Reductions in free available chlorine concentration of 12% and 10% were observed at 13°C in the 28-day holding period. Note that no significant change in concentration was observed in the 5% solution at 13°C after the 28-day period. As expected, data at 27°C show a higher reduction in free available chlorine concentration for the 15 and 10% free available chlorine solutions. Although reductions in free available chlorine concentrations as high as 28% were observed for the 15% solution at 27°C after 28 days, no significant change was observed for the 5% sodium hypochlorite solution.

Table 3.1 Free Available Chlorine and Chlorate Ion Concentrations in Commercial Strength Sodium Hypochlorite held in the dark for 28 days.

Commercial Hypochlorite Free Available Chlorine (%)	Temperature (°C)	Measured Free Available Chlorine (g/l)	Reduction in Free Available Chlorine (%)	Measured ClO_3^- Concentration (g/l)
15	13	132.6	12	2.99
10		90.1	10	1.17
5		53.5	N/C	0.50
15	27	108.1	28	12.71
10		83.5	17	4.56
5		50.1	N/C	1.07

Note: N/C = No Change

Source: Reprinted from Journal AWWA, Vol. 89, No. 4 (April 1997), by permission. Copyright American Water Works Association.

Typically, commercial grade sodium hypochlorite is held for 14 days at a concentration of approximately 15%. On-site generated sodium hypochlorite is typically stored for only 1 day at a concentration of 5 to 8% providing for a very stable free available chlorine concentration. Based on the above data, it can be reasoned that if sodium hypochlorite is to be held for extended periods of time, it should be held at lower strength concentrations (e.g., 5% free available chlorine concentrations).

Reductions in concentration of sodium hypochlorite in the storage tank may make it difficult for some operators to apply the correct chlorine dose. An incorrect chlorine dose may result in an insufficient residual being maintained in a drinking water distribution system (a health risk) or an excess of chlorine being discharged from a wastewater treatment plant into a receiving water (an environmental quality issue).

Gordon et al. (1995b) performed preliminary metal ion experiments to obtain general information concerning the reduction in free available chlorine in stored commercial sodium hypochlorite in the presence of individual transition metals. In these experiments 1 mg/l of Ni^{2+}, Cu^{2+}, Mn^{2+}, and Fe^{3+} were separately added to commercial sodium hypochlorite solutions, stored in the dark at 25°C, and periodically analyzed during a 60-day period. The sodium hypochlorite decomposition results were compared to a sodium hypochlorite blank containing no transition metals.

These preliminary experiments showed that addition of 1 mg/l of Cu^{2+} accelerated sodium hypochlorite decomposition by a factor of 1.4 relative to the blank. The addition of 1 mg/l of Ni^{2+} was observed to greatly enhance the rate of sodium hypochlorite decomposition. The addition of 1 mg/l Mn^{2+} and Fe^{3+} resulted in no increase in the rate of sodium hypochlorite decomposition relative to the blank. These data indicate the necessity of specifying low transition metals concentrations in commercial strength sodium hypochlorite.

Gordon et al. (1995a) reported that the role of transition metals in the decomposition of sodium hypochlorite is complex. However, the maximum concentration of transition metals should be limited to less than 0.1 mg/l of Ni^{2+} and 1 mg/l Cu^{2+} in stored sodium hypochlorite. Ferric iron and manganese, when present alone, are effective catalysts for sodium hypochlorite decomposition. It should be noted that on-site generated sodium hypochlorite is less susceptible to transition metal effects due to its lower initial concentration and the short time from production to application.

The loss of strength in sodium hypochlorite solutions may also result in the formation of undesirable by-products (e.g., chlorate) as shown in the following equation (Bolyard et al., 1992).

$$3\ ClO^- = 2Cl^- + ClO_3^-$$

Bolyard et al. (1992; 1993) reported that the mass concentration of chlorate ranges from 1.7 to 220% of the mass concentration of free available chlorine. However, the concentration of chlorate in stock solutions is a function of solution strength, aging time, temperature, pH, and the presence of metal catalysts (Gordon et al., 1993; 1995). Gordon et al. (1997) also presented data concerning chlorate ion formation in commercial sodium hypochlorite held in the dark in glass containers for 28 days (see Table 3.1). As expected, these data indicate that increasing chlorate ion formation was observed with increasing sodium hypochlorite strength. The most dramatic increase in chlorate ion formation is observed with increasing the temperature of the sodium hypochlorite during the 28-day holding period. When comparing the 27°C data with the 13°C data the following was observed after the 28-day holding period (see Table 3.1):

- A 425% increase in chlorate ion concentration for the 15% solution
- A 390% increase in chlorate ion concentration for the 10% solution
- A 214% increase in chlorate ion concentration for the 5% solution

Again, as with the stability information (discussed above) less chlorate ion is formed in lower strength sodium hypochlorite (e.g., 5% free available chlorine) than in commercial strength sodium hypochlorite regardless of temperature.

Presently, chlorate ion is not regulated in the current USEPA primary drinking water regulations. However, as of January 1, 2002, chlorite ion is regulated for systems using chlorine dioxide as a disinfectant. The maximum contaminant level for chlorite is 1.0 mg/l and the maximum contaminant level goal is 0.8 mg/l. It should be noted that chlorate ion is not listed in the USEPA candidate contaminant list (CCL) for drinking water. However, perchlorate is listed in the CCL.

Gordon et al. (1997) developed a computer-based sodium hypochlorite decomposition model (the Gordon-Adam model) that was also shown to be a good predictor of chlorate ion formation in stored sodium hypochlorite.

chapter 3: Disinfection applications and alternatives   13

Following extensive testing of the model, Gordon recommended three basic strategies for minimizing chlorate ion concentration and increasing the stability of commercial grade sodium hypochlorite as follows:

- Dilute concentrated sodium hypochlorite immediately after delivery.
- Use lower storage temperatures.
- Avoid sunlight during storage

3.3 Ozone

Ozone is a very effective disinfectant for both water and wastewater treatment applications. It has been shown to be excellent for the disinfection of pathogens of concern in both water treatment distribution systems and wastewater treatment outfalls (e.g., *Giardia* and *Cryptosporidium*). It also does not readily contribute to the formation of trihalomethanes in drinking water treatment and distribution systems or wastewater treatment outfalls. However, the primary drawback to ozone applications in drinking water treatment systems is its lack of an ability to provide a disinfectant residual for the distribution system. It is also relatively more expensive than other disinfectants discussed in this chapter.

3.4 Chlorine dioxide

Although chlorine dioxide was first produced in 1811, it was not until the middle 1960s when its widespread use occurred due to changes in the manufacturing process. As with ozone, chlorine dioxide is an effective disinfectant in both water and wastewater treatment systems. It also does not have a tendency to form trihalomethanes in water treatment distribution systems or wastewater treatment outfalls. Chlorine dioxide applications have typically been associated with medium-sized treatment facilities in the United States.

References

Baker, R.J., (1969), "Characteristics of Chlorine Compounds," *Journal of the Water Pollution Control Federation*, 41:482.

Boylard, M., Fair, P.S., and Hautman, D.P., (1992), "Occurrence of Chlorate in Hypochlorite Solutions Used for Drinking Water Disinfection," *Environmental Science and Technology*, 26(8):1663-1665.

Boylard, M., Fair, P.S., and Hautman, D., (1993), "Sources of Chlorate Ion in US Drinking Water," Journal AWWA, 85(9):81-88.

Gordon, G.L., Adam, C., Bubnis, B.P., Hyot, B., Gillette, S.J., and Wilczak, A., (1993), "Controlling the Formation of Chlorate Ion in Liquid Hypochlorite Feedstocks," *Journal AWWA*, 85(9):89-97.

Gordon, G., Adam, L., and Bubnis, B., (1995a), *Minimizing Chlorate Ion Formation in Drinking Water When Hypochlorite Ion is the Chlorinating Agent*, American Water Works Association Research Foundation (Order Number 90675).

Gordon, G., Adam, L., Bubnis, B., (1995b), "Minimizing Chlorate Ion Formation in Drinking Water When Hypochlorite Ion is the Chlorinating Agent, *Journal of the American Water Works Association*, 87:6, 97-106.

Gordon, G., Adam, L., Bubnis, B., Kuo, C. Cushing, R., and Sakaji, R., (1997), "Predicting Liquid Bleach Decomposition," *Journal of the American Water Works Association*, 89:4, 142-149.

Laubusch, E., (1963), "Sulfur Dioxide," *Public Works*, 94(8):117.

chapter 4

Disinfection chemistry

4.1 Chlorine application chemistry

4.1.1 Basic principles of disinfection

Disinfection in water and wastewater treatment systems may be defined as the destruction of pathogens (e.g., bacteria, viruses, protozoan, or amoebic cysts) to provide public health protection. Disinfectant chemicals for use in water, wastewater, and cooling tower applications include chlorine and chlorine compounds (elemental chlorine, chloramines, sodium hypochlorite, calcium hypochlorite, and chlorine dioxide), ozone, UV radiation, and a variety of other physical and chemical agents.

Design considerations and operational factors for disinfectant systems are as follows: the microorganisms to be inactivated; the concentration of the microorganisms in the water; the water quality in which disinfection will occur; the variability of the water quality; the type of disinfectant chosen for application; the dose or concentration of the disinfectant applied; and the contact time of the disinfectant with the water.

Generally, the adequacy of disinfection in water and wastewater treatment systems is determined by the product CT, where C is the final residual concentration of the disinfectant in the water and T is the contact time that is exceeded by 90% of the fluid (Haas, 1999). CT values necessary to achieve varying levels of disinfection in water treatment systems have been published in USEPA guidance documents for chlorine and chlorine compounds, ozone, chlorine dioxide, and chloramines (Malcolm Pirnie and HDR Engineering, 1991). It should be noted that CT values published in the above referenced document also include an acceptable margin of safety for design purposes.

4.1.2 Chemistry of elemental chlorine and sodium hypochlorite in water

Chlorine is used in water and wastewater treatment processes as both a disinfectant and an oxidizing agent. Chlorine may be applied as gaseous chlorine that is dissolved in water, liquid sodium hypochlorite, or solid calcium hypochlorite. All three forms of chlorine application are chemically equivalent due to the rapid equilibrium between each of the forms when they are injected into water.

Elemental chlorine (Cl_2) is a dense gas that condenses to a liquid when subjected to pressures in excess of its vapor pressure. Note that cylinder pressure at sea level is approximately 160 lbs/in² (psi). As a result, commercial shipments of chlorine are made in pressurized tanks to reduce shipment volume. When the pressurized chlorine liquid is dispensed, the pressure is reduced (via a vaporizer) to vaporize the compressed liquid chlorine into a gas for application. The dissolution of gaseous chlorine in water forms aqueous chlorine. This reaction may be expressed by Henry's law and the corresponding Henry's Law Constant (H).

$$Cl_{2(g)} = Cl_{2(aq)}$$

$$H \text{ (mol/ l–atm)} = [Cl_2(aq)]/P_{Cl2}$$

where H is Henry's Law Constant, [Cl_2(aq)] is the molar concentration of aqueous chlorine, and P_{Cl2} is the partial pressure of chlorine in the gas phase in equilibrium with the liquid (measured in atmospheres).

When aqueous chlorine is combined with water it rapidly forms hypochlorous acid, protons, and chloride ions according to the following equation.

$$Cl_{2(aq)} + H_2O = HOCl + H^+ + Cl^-$$

Hypochlorous (HOCl) acid is a relatively weak monoprotic acid that may dissociate according the following equation.

$$HOCL = OCl^- + H^+$$

where $pK_A = 7.54$ at 25°C (see Figure 4.1).

Free available chlorine is a term used to refer to the concentrations of molecular chlorine (Cl_2), hypochlorous acid (HOCl), and hypochlorite ion (OCl^-) in water expressed as available chlorine. The term available chlorine is used to express the relative amount of chlorine present in chlorine gas or hypochlorite salts. Available chlorine is expressed by determining the electrochemical equivalent amount of Cl_2 to the compound present in water (Haas, 1999).

The distribution of free chlorine in water between hypochlorous acid and hypochlorite ion as a function of pH is shown in Figure 4.1. Note that at very high pH values, almost all the free chlorine exists as hypochlorite ion. At very low pH values, almost all the free chlorine exists as hypochlorous acid. Equal quantities of hypochlorous acid and hypochlorite ion exist at the pK_A value of 7.54.

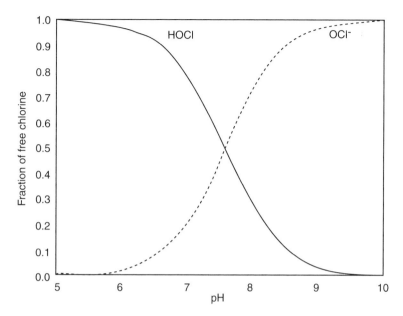

Figure 4.1 Effect of pH on the fraction of hypochlorous acid and hypochlorite ion at 20°C. (From Haas, C.N., *Water Quality and Treatment: A Handbook of Community Water Supplies*, AWWA, McGraw-Hill, New York, 1999. With permission.)

Based on the equation shown below, one mole of elemental chlorine is capable of reacting with two electrons to form chloride:

$$Cl_2 + 2e^- = 2\ Cl^-$$

Since one mole of hypochlorite (OCl$^-$) also reacts with two electrons to form chloride (as shown below), one mole of hypochlorite is electrochemically equivalent to one mole of elemental chlorine. Thus, one mole of elemental chlorine and one mole of hypochlorite both contain approximately 71 g of available chlorine (Haas, 1999).

$$OCl^- + 2e^- + 2H^+ = Cl^- + H_2O$$

4.1.3 Chlorine demand reactions

The amount of chlorine added to water for disinfection purposes is termed the dose and is usually reported in pounds of chemical applied per day. Some fraction of the applied disinfectant will react with substances (both organic and inorganic) in the water, causing a reduction in the amount of disinfectant remaining in the water (this disinfectant reduction is termed the demand). The amount of disinfectant remaining in the system after a set time period is called the residual and usually reported in mg/l or ppm (parts per million). The following equation expresses the relationship discussed above:

$$\text{Dose} = \text{Demand} + \text{Residual}$$

Chlorine demand reactions will be divided into two sections for purposes of discussion. First, chlorine reactions with ammonia and their significance to disinfection practice will be presented and discussed. Next, chlorine reactions with organic and inorganic matter will be presented and discussed.

4.1.3.1 Demand reactions with ammonia

When free chlorine is added to water in the presence of ammonia, it reacts in a stepwise manner to form monochloramine (NH_2Cl), dichloramine ($NHCl_2$), and trichloramine (NCl_3), which is otherwise known as nitrogen trichloride (Davis and Cornwell, 1998).

$$NH_4^+ + HOCl = NH_2Cl + H_2O + H^+$$

monochloramine

$$NH_2Cl + HOCl = NHCl_2 + H_2O$$

dichloramine

$$NHCl_2 + HOCl = NCl_3 + H_2O$$

trichloramine or nitrogen trichloride

The sum of these three compounds is usually defined as the combined chlorine residual in water. The distribution of the species among the three chloramine species is a function of time, temperature, pH, and the initial $Cl_2:NH_4$-N ratio.

Figure 4.2 is a schematic of a breakpoint chlorination curve. This schematic may be divided into three zones. Zone 1 is where all chlorine added to the water results in the formation of chloramines. This zone occurs at chlorine doses below a $Cl_2:NH_4$-N ratio of 5. During this zone there is no change in the nitrogen concentration in the water. Chloramines are oxidized in zone 2, resulting in a loss of nitrogen occurring in the water also in zone 2. The breakpoint occurs at a $Cl_2:NH_4$-N ratio of 7.6 where all of the chloramines have been oxidized and zone 3 begins. In zone 3, all chlorine added to the water exists as free chlorine. The breakpoint indicates the amount of chlorine that must be added to a water containing ammonia before a stable free chlorine residual can be obtained.

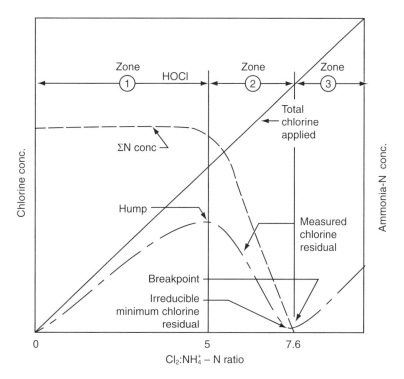

Figure 4.2 Idealized schematic of a breakpoint chlorination curve. (From White, G.C., *Handbook of Chlorination and Alternative Disinfectants*, Copyright © 1999. This material is used by permission of John Wiley & Sons, Inc.)

4.1.3.2 Demand reactions with organic and inorganic matter

Chlorine can react with both inorganic and organic material to form a variety of substances. Granstrom and Lee (1957) observed that phenol could be chlorinated by free chlorine to form chlorophenols. Morris (1967) determined that organic amines react with free chlorine to form organic monochloramines. Murphy (1975) indicated that phenols, amines, aldehydes, ketones, and pyrrole groups are susceptible to chlorination. Wojtowicz (1979) summarized the reaction rates for free chorine residuals and inorganic compounds. More recently, Krasner (1986) determined that free chlorine was capable of removing tastes and odors associated with organic sulfur compounds.

4.1.4 Disinfection kinetics

Microorganism inactivation during disinfection is based upon the initial work of Chick (1908) who developed the relationship between microbial inactivation by chemical disinfectants and chemical reactions as shown below:

$$r = -kN$$

where k is the reaction rate constant, r is the inactivation rate, and N is the concentration of viable microorganisms.

Watson (1908) proposed an equation that related the rate constant of inactivation (k) to the disinfectant concentration (C). The combination of the concepts from these two equations resulted in the classical formulation of the Chick-Watson law shown below:

$$\ln(N/N_o) = k'C^n t$$

where N and N_o are the concentrations of viable microorganisms at time t and time zero, respectively. Other terms in the equation are defined as shown above.

In 1972, Hom developed a modification of the Chick-Watson law in the following form;

$$r = -k'mt^{m-1}NC^n$$

Following integration and simplification, the Hom model takes on the following form:

$$\ln(N/N_o) = = k'C^n t^m$$

A detailed overview of the principles of disinfection modeling was presented by Gyurek and Finch (1998).

4.2 On-site sodium hypochlorite generation chemistry

The objective of an electrochemical process is to control the conditions in an electrolytic cell so that the desired products are generated. An electrochemical process requires three basic components as shown below:

- A source of direct electrical current (usually a rectifier connected to an AC power source)
- An electrolytic cell
- Electrical conductors

Although the electrolytic cell has many proprietary components, it can be divided into the following components necessary for an electrochemical process to occur:

- A cell body to contain the cell components
- Electrodes
  - Anode — positive electrode – electron deficient
  - Cathode — negative electrode – electron rich
- Electrolyte (water plus an acid, base or salt)
- Electrical connectors

4.2.1 Electrolytic cell reactions

Once power is provided to the cell, electrons move from the anode to the cathode. This electron transfer results in the development of an electrical potential between the two electrodes. Several chemical reactions may occur singularly or in combination at the electrodes as listed below:

- Dissolution of the anode
- Oxidation of an anion at the anode
- Reduction of a cation at the cathode by the following reactions
  - Deposition of a metal ion
  - Reduction of water to hydrogen
  - Forcing an atom or molecule to accept one of more electrons without either deposition of a metal ion or reduction of water to generate hydrogen

The net reaction in the hypochlorite cell to produce sodium hypochlorite is shown below where sodium chloride and water produces sodium hypochlorite and hydrogen gas:

$$NaCl + H_2O = NaOCl + H_2$$

The amount of sodium hypochlorite produced is a function of the amount of direct electrical current passed through the electrolytic cell. The above reaction can be divided into the principal reactions at the anode. In this reaction, chloride ion is oxidized as shown below:

$$2Cl^- - 2e = Cl_2$$

The principal reactions in the solution occur after the chloride generated at the anode is rapidly hydrolyzed into hypochlorous acid.

$$Cl_2 + 2H_2O = HOCl + H^+ + Cl^-$$

The principal reaction in the hypochlorite cell at the cathode is:

$$2H_2O + 2e = H_2 + 2OH^-$$

$$HOCl = H^+ + OCl^-$$

In addition to the principal reactions in the electrolytic cell, there are also parasitic reactions occurring in the cell that reduce the amount of hypochlorite produced in the system. The parasitic reactions at the anode are shown below:

$$2H_2O = O_2 + 4H^+ + 4e \text{ (oxygen producing)}$$

$$6OCl^- + 3H_2O = 2ClO_3^- + 6H^+ + 4Cl^- + 1.5 O_2 + 6e \text{ (oxygen producing)}$$

$$ClO^- + H_2O + 2e = 2H^+ + Cl^- + O_2 \text{ (oxygen producing)}$$

The parasitic reaction occurring at the cathode is as follows:

$$OCl^- + H_2O + 2e = Cl^- + 2OH^-$$

The parasitic reaction in the solution is:

$$2HOCl + OCl^- = ClO_3 + 2H^+ + 2Cl^-$$

In addition to the principal reactions and the parasitic reactions occurring to a hypochlorite cell as shown above, an additional oxygen producing reaction may also occur as shown below:

$$2ClO^- = O_2 + 2Cl^-$$

Water hardness can also form a scale on the cathode of the cell. The reactions for scale formation are as follows:

$$HCO_3^- + NaOH = CO_3^{2-} + H_2O + Na^+$$

Once generated, the carbonate ion reacts with the calcium present in the feed water to form calcium carbonate according to the following chemical equation.

$$Ca^+ + CO_3^{2-} = CaCO_3$$

This calcium carbonate forms a scale that may adhere to the cathode depending upon the water quality conditions in the cell. These and other deleterious reactions in the cell, their impact on system, system design, and operation will be discussed in subsequent chapters.

References

Chick, H., (1908), "An Investigation of the Laws of Disinfection," *Journal of Hygiene*, 8:92-157.

Davis, M.L. and Cornwell, D.A., (1998), *Introduction to Environmental Engineering*, McGraw-Hill, 3rd ed., New York.

Fair, G.M., Morris, J.C., Chang, S. L., Weil, I., and Burden, R.P., (1948), "The Behavior of Chlorine as a Water Disinfectant," *Journal of the American Water Works Association*, 40:1051-1061.

Gordon, G., Adam, L., and Bubnis, B., (1995a), *Minimizing Chlorate Ion Formation in Drinking Water When Hypochlorite Ion is the Chlorinating Agent*, American Water Works Association Research Foundation (Order Number 90675).

Gordon, G., Adam, L., and Bubnis, B., (1995b), "Minimizing Chlorate Ion Formation in Drinking Water When Hypochlorite Ion is the Chlorinating Agent, *Journal of the American Water Works Association*, 87:6, 97-106.

Gordon, G., Adam, L., Bubnis, B., Kuo, C. Cushing, R., and Sakaji, R., (1997), "Predicting Liquid Bleach Decomposition," *Journal of the American Water Works Association*, 89:4, 142-149.

Gordon, G., (1998), "Electrochemical Mixed Oxidant Treatment: Chemical Detail of Electrolyzed Salt Brine Technology," USEPA, May, 1998.

Granstrom, M.L. and Lee, G.F., (1957), "Rates and Mechanisms of Reactions Involving Oxychlorine Compounds," *Applied Environmental Microbiology*, 46:619.

Gyurek, L.L. and Finch, G.R., (1998), "Modeling Water Treatment Chemical Disinfection Kinetics," *Journal of Environmental Engineering*, 124(9), 783-793.

Haas, C.N., (1999), "Disinfection," *Water Quality and Treatment: A Handbook of Community Water Supplies*, AWWA, McGraw-Hill, Letterman, R.D., Ed., New York.

Haas, C.N. and Karra, S.B., (1984), "Kinetics of Microbial Inactivation by Chlorine. II. Kinetics in the Presence of Chlorine Demand," *Water Research*, 18:1451-1454.

Hoff, J., (1986), "Inactivation of Microbial Agents by Chemical Disinfectants," EPA/600/2-86/067, USEPA.

Hom, L.W., (1972), "Kinetics of Chlorine Disinfection of an Ecosystem," *Journal of Sanitary Engineering Division*, ASCE, 98(SA1):183-194.

Krasner, S., (1996), Presented at the AWWA Annual Conference, Denver, CO.

Malcolm Pirnie and HDR Engineering, (1991), *Guidance Manual for Compliance with Filtration and Disinfection Requirements for Public Water Systems Using Surface Water Sources*, American Water Works Association.

Morris, J., (1967), "Kinetics of Reactions Between Aqueous Chlorine and Nitrogen Compounds," *Principles and Applications of Water Chemistry*, S. Faust, Ed., John Wiley & Sons, New York.

Murphy, K., (1975), "Effect of Chlorination Practice on Soluble Organics," *Water Research*, 9:389.

Oliveri, V. et al., (1971), "Inactivation of Virus in Sewage," *ASCE Journal of Sanitary Engineering Division*, 97(5), 661.

Sodium Hypochlorite Manual, The Chlorine Institute, Inc., Pamphlet 96, Edition 2, May, 2000.

Sodium Hypochlorite Stability Information Required for EPA Labels and Chlorine Institute Stability Test Data, The Chlorine Institute, Inc., Member Information Report, 143.

Trumm, J., (1994), "SCOLA Characterization: An Analysis of the Products of Brine Electrolysis," M.S. thesis, The University of New Mexico, December, 1994.

Venczel, L.V., Arrowood, M., Hurd, M., and Sobsey, M.D., (1997), "Inactivation of *Cryptosporidium* parvum oocysts and Clostridium perfringes spores by a mixed-oxidant disinfectant and by free chlorine," *Applied Environmental Microbiology*, 63, 4, 1598-1601.

Watson, H.E., (1908), "A Note on the Variation of the Rate of Disinfection with Change in the Concentration of the Disinfectant," *Journal of Hygiene*, 8:536-542.

White, G.C., (1999), *Handbook of Chlorination and Alternative Disinfectants*, John Wiley & Sons, New York.

Wojtowicz, J.A., (1979), "Chlorine Monoxide, Hypochlorous Acid and Hypochlorites," *Kirk-Othmer Encyclopedia of Chemical Technology*, 3rd ed., vol. 5, John Wiley & Sons, New York, 580-611.

chapter 5

Electrolyzer systems

5.1   Electrolyzer system types and principles of operation

Electrolyzer systems can be classified into two basic types, brine electrolysis and seawater electrolysis. The basis for classification is the feedstock derived from either crystallized salt for brine systems or seawater feed for seawater electrolysis systems.

Although the product of each system is the same sodium hypochlorite disinfectant, differences in the electrolysis method exist as a result of the variations in the calcarious hardness and other properties of the feed material. Since crystallized salt is dissolved and used for electrolysis in brine systems, control of the calcarious components may be achieved using water softening or selecting the desired quality of the crystallized salt. Seawater does not allow for easy methods of calcarious component control. Thus, an entirely different approach to electrolysis is used in seawater systems.

To electrolyze a brine solution for sodium hypochlorite production, brine electrolysis cells are designed for very low brine feed flow rates with narrow electrode gaps that produce sodium hypochlorite concentrations approaching 1%. The seawater system approach is to use very high seawater flow rates with wide electrode gaps that produce sodium hypochlorite concentrations of less than 0.3% to reduce the rate of deposit formation on the cathodes. Brine systems have an average current efficiency of 65% while seawater systems have an average current efficiency greater than 80%. This difference in current efficiency has an effect on power consumption and, for the brine system, on salt consumption.

5.2   Brine system: general description

Brine systems can be used for any application requiring chorine or chloramines as a part of the disinfection regimen. These systems are nearly always installed inland and are designed to provide substantial quantities of stored sodium hypochlorite. Brine systems are designed with excess product storage to assure that disinfection capacity is always available to the end user.

To accommodate these requirements systems are generally configured with the following components and operate in the manner described below:

- Water softener: Essential for removal of calcium and magnesium from the feed water
- Salt dissolver: Provides the required salt solution for electrolysis
- Electrolyzer cell or cells: Electrolyzes the dilute brine solution
- DC power rectifier: Provides the direct current for electrolysis
- Storage tanks: Product storage to meet dosing requirements as well as any excess capacity essential to assure continuous dosing capabilities
- Hydrogen dilution blowers: Provided to dilute the by-product hydrogen produced during the electrolysis process
- Dosing pumps with dosing controls: Provide the needed disinfection dose based upon the chlorine residual or flow rate at the point of disinfection
- Cell cleaning system: Used to remove the calcarious material deposited on the cell cathodes during the production process
- Central control panel: Performs the system production control function

A simple brine system operation begins with the client's domestic water supply flowing through the water softener where the water hardness is reduced. A portion of the softened water is added to the salt dissolver to make a concentrated salt brine solution of approximately 300 g/l. The concentrated salt brine is then mixed with the main stream of softened water to produce a final brine concentration of approximately 3% (30 g/l) salt concentration. This final brine solution is then pumped through the electrolyzer cell. The cell electrolyzes the final brine solution into sodium hypochlorite that is then forced by the incoming water pressure to flow to the storage tank. Sodium hypochlorite from the storage tank is used as a supply source for dosing pumps. These pumps are operationally controlled by either a residual analysis or a flow-pacing signal to supply sodium hypochlorite to the point of disinfection application. Because the calcarious material is controlled in this system, electrolyzer cell cleaning frequency varies from 1 month to 6 months. Cell cleaning is accomplished using either hydrochloric or sulfamic acid at a concentration ranging from 5 to 10%. Note, raw water hardness should always be less than 50 mg/l as calcium carbonate in the cell feed to control monthly cathode acid cleaning.

Brine systems are susceptible to water temperatures less than 15°C. If water temperatures are less than 15°C, the water temperature must be raised by adding some form of heat exchanger to the outlet of the cell. The inlet feed water is passed through one side of the exchanger and heated product through the opposing side. Exchangers are sized to assure that the system inlet water is heated to at least 15°C. Heat exchangers normally have a plastic or stainless steel shell with titanium tubes, tube sheets, and heads or, alternatively, plate exchangers of titanium with EPDM or Viton sealing materials.

5.3 Seawater system: general description

Seawater systems are, as the name implies, applicable only in coastal areas. These systems are applied to control biological activity in all forms of circulating cooling systems where seawater is the cooling media such as power stations, refinery complexes, fertilizer plants, etc. Many of these applications are remote from other sources of disinfection, making seawater electrochlorination very cost effective.

Seawater electrochlorinators utilize high seawater feed flow rates to help control cathode fouling from the naturally occurring magnesium and calcium in seawater. As a result, the product concentration of sodium hypochlorite is low and operating current efficiency is high. Because seawater provides a "free" source of salt, these systems contain smaller equipment of less variety than brine electrolysis systems. The equipment for a seawater system is shown below, and generally operates in the manner described:

- Inlet seawater strainers: Essential for removing particulate material from the seawater feed stream
- Seawater booster pumps: Provide seawater at the appropriate pressure and flow to the electrolyzer system
- Electrolyzer cell or cells: Electrolyzes the seawater feed to sodium hypochlorite
- DC power rectifier: Provides the direct current for electrolysis
- Hydrogen degassing storage tanks: For removing by-product hydrogen and product storage to meet shock dosing requirements
- Hydrogen degassing cyclone: An alternate degassing method where only continuous disinfectant dosing is required and hydrogen removal is desirable
- Hydrogen gas seal pot: Used with the cyclone where hydrogen is vented to the atmosphere without dilution
- Hydrogen dilution blowers: Provided to dilute the by-product hydrogen produced during the electrolysis process. These blowers may be applied to either tank or cyclone hydrogen removal systems
- Dosing pumps: Provide the needed disinfection dose to distant or high pressure applications. Direct current variation controls the dose rate rather than control of the pump flow
- Cell cleaning system: Used to remove the calcarious material deposited on the cathode

Seawater systems are less complex than brine systems; there are no water softening or brine dissolving requirements. Seawater is drawn from a source and pumped through seawater strainers to remove larger suspended material. From the strainers seawater passes through a flow controller and flow meter to the electrolyzer cells where direct current converts the solution to sodium hypochlorite. The product then enters either a storage tank or a cyclone assembly for hydrogen removal. Hydrogen removed from the liquid stream is diluted by air fans to 2% or less, then vented to atmosphere. Where a storage tank is used, the dosing pumps operate from level controls for either continuous, a combination of continuous and shock dosing, or shock dosing only. Dosing rate control is provided by adjusting the direct current applied to the eletrolyzer cells while maintaining a stable seawater flow to the cells.

Seawater cells are designed with both a wide electrode gap and very high flow velocities to control deposit formation, always a problem in seawater electrolysis. Seawater can have wide variations in deposit causing hardness in different ocean locations. For example, hardness found in Arabian Gulf seawater is approximately 500 mg/l calcium and 1500 mg/l magnesium. However, seawater around Taiwan has 400 mg/l calcium and 1250 mg/l magnesium.

Another issue for seawater cells is operation in low salinity water conditions. Salinity is classified as 18,980 mg/l chloride at 100% seawater conditions. Systems designed for operations at chloride conditions below 80 % (15,185 mg/l chloride) must be sized for reduced output performance as empirically derived by the individual cell manufacturer. Anode coatings designed for elevated oxygen conditions have been built and operated in salinity levels down to 2500 mg/l chloride. Note, it should be understood that the quantity of cells required to meet a specific need must be increased to assure adequate sodium hypochlorite production.

As with the brine systems discussed above, similar conditions also exist for cells in operation where water temperatures are below 15ºC. As noted, both the electrode coatings used and the number of cells employed in a system must be adjusted to assure adequate system performance.

5.4 Electrolytic cells

Electrolytic cells capable of producing sodium hypochlorite have been in existence on a laboratory scale for well over 100 years. Cell equipment before the invention of dimensionally stable anodes (DSA) was very inefficient, cumbersome, and costly. DSA development simplified cell designs to allow cost efficient production of sodium hypochlorite on-site.

Electrolytic cells may be designed having monopolar or bipolar electrode configurations. Each of these design configurations is discussed below.

A monopolar cell consists of an anode (direct current positively charged member) and cathode (direct current negatively charged member) each joined to the power source by a separate power connection. These electrodes are separated by a space that allows the salt solution to flow between the plates for electrolysis to occur. Multiple electrodes may be connected to a common input connection in a parallel configuration within each electrolyzer assembly. Each electrode set polarity is defined by its connection polarity, anodes only on the positive connection and cathodes only on the negative connection.

chapter 5: Electrolyzer systems

Bipolar electrodes differ from those discussed above in that each electrode will serve as both an anode and a cathode. The bipolar cell design will have terminal electrodes for the positive and negative power input points and interstitial bipolar electrodes. Direct current is delivered to the positive DSA coated terminal electrode face, emitted from that electrode face through the brine solution, then is received on the cathode face of the adjacent plate and passes through the plate to the anode face of the same electrode. Each electrode has a DSA coated portion and a non-DSA coated portion. Current flow through the cell proceeds alternately through each bipolar electrode set in the cell to the non-DSA coated cathodic terminal electrode.

Cell designs are divided into many basic categories as follows (Table 5.1):

A. Flat plate type bipolar cells that utilize a filter press configuration wherein one face of each electrode is anodic and the other face cathodic
B. Flat plate bipolar arrangements in an FRP, PVC or acrylic tube/pipe having individual compartments within the cell and a terminal electrode set for each DC power connection
C. Flat plate monopolar arrangements in a rectangular rubber lined steel, FRP, PVC, or polypropylene body having electrode connections from one side to allow connection to both the positive and negative DC power connections
D. Tubular bipolar arrangements of titanium. In this cell, the inner tube is bipolar and the outer tubes are monopolar. The bipolar tube is half coated, half non-coated. An opposite polarity monopolar outer tube mate to each bipolar section is where the DC power is connected to the cell.

Cell designs are defined by their application because operating conditions vary widely between brine and seawater electrolysis. Table 5.1 summarizes electrolytic cell manufactures, the type of electrolytic cell design classification, and the general equipment application.

5.4.1 Tubular and plate electrolytic cell designs

Electrolyzer cells can be divided into two basic design categories, plate electrode designs and tubular electrode designs. The major variation on these

Table 5.1 Electrolytic Cell Manufacturers

Company	Design	Electrolysis Use
Japan Carlit	A	Brine systems
Severn Trent Services – ClorTec	B, D	Brine systems
Severn Trent DeNora	B, C, D	Seawater systems
USFilter Electrocatalytic	D	Seawater systems
USFilter Wallace and Tiernan	B	Brine systems
Pepcon, USA	B	Seawater systems
Mitsubishi, Japan	C	Seawater systems
Daiki	B	Seawater systems

two designs is the tubular cell with plate electrodes. Table 5.2 summarizes the electrolytic cell configurations used by various manufacturers.

Cell performance is defined as the power efficiency to produce 1 lb (kg) of chlorine equivalent. This performance is controlled by several variables: seawater salt content, seawater temperature, seawater flow rate, anode coating, cathode type, cell current density, and the number of cells connected in series hydraulically. Because most cells today use very similar anode coatings and cathode types, the dominant variables affecting performance, in order of importance, are: number of cells in series, cell operating current density, seawater salt content, and seawater temperature.

As the number of cells in series is increased, the product concentration increases. At concentrations above 2.0 g/l (0.2%) operating current efficiency falls below 80% regardless of the variable combination employed. This reduction in efficiency translates to increased power consumption.

A high product concentration does not increase disinfection performance nor does it improve the system cost structure. Generally, the most power efficient system will provide low product concentrations, less than 1 g/l (0.1%), and support disinfection needs very effectively.

5.4.1.1 ELCAT — Chloromat

A tubular cell arrangement was patented in 1972, U.S. Patent 1,396,019, by Engelhard Minerals and Chemicals, Ltd. and subsequently sold to a group of investors for the Electrocatalytic Corporation, which was subsequently purchased in 2000 by USFilter Corporation. The tube inside a tube configuration is very successful in providing reliable seawater electrolysis in a 1 lb/h (24 lb/d) per cell format.

Since the cell output is small, many cells may be required to produce a large quantity of sodium hypochlorite. This system may not be suitable for large hypochlorite capacity requirements. For example, a plant requiring 1 metric ton of chlorine equivalent must have 92 cells to meet only specific needs without regard for back-up capacity to assure that chlorine is always available for system disinfection.

This cell is a bipolar design where the inner tube is the bipolar element. The terminal electrodes are the mono-polar outer tubes. Electrode spacing is controlled by a centering arrangement at the cell ends. Uniform electrode spacing and seawater flow through the cell controls fluid velocity in the annulus between the electrodes, reducing the requirement for cathode cleaning.

Cathode cleaning is a necessary part of any seawater electrolyzer maintenance program. An alkaline, sodium hydroxide, layer is created on the cathode surface during electrolysis reacting with the magnesium and calcium in the seawater to cause cathodic surface deposits (primarily magnesium hydroxide, calcium hydroxide, and calcium carbonate). It is important to note that cell cleaning is required regardless of the electrolytic cell design. Acid cleaning is essential to reliability of the electrolyzer cell. The materials used for this procedure are hydrochloric and sulfamic acids. Acid concentrations vary with manufacturer; however, a concentration range of 5 to 10% is generally acceptable for all cell designs.

Originally developed in England, once patented in the United States, the tubular cell was very successfully marketed to consulting engineering firms for offshore platform cooling and firewater systems, ocean going ship cooling systems, and power station open loop cooling systems.

Tubular electrolyzer arrangements are normally a horizontally mounted series of cells using multiple hydraulically and electrically parallel rack mounted trains attached to a single rectifier. The seawater supply and sodium hypochlorite headers are mounted at the front of the module assembly forming a U-shaped loop on each parallel train to each header. Normally the number of cells in series is limited to 12 per train. This cell design is able to withstand test pressures of 150 psig offering some advantages in specific applications by occasionally eliminating the need for dosing pumps.

Electrical connections are made where the positive connection is made to the middle of the loop and the negative connection is made on the inlet and outlet. The DC current path is then parallel down each side of the loop while the inlet and outlet are neutral to ground. Although this configuration helps prevent current leakage into the surrounding piping systems, this electrical arrangement also creates a problem for balancing the parallel electrical circuits. Since the cells are another form of electrical resistor, a resistance change in any parallel cell segment will cause the current to increase in the segment with lower resistance or vice versa, in turn causing reduced anode life and increased deposit formation due to the higher current.

Claims are made that these cells require no cleaning. However, all seawater cells must have routine cleaning and maintenance. Although the very high fluid velocities in this cell tend to scrub deposit formations, it will not entirely eliminate deposit formation. A cleaning frequency of at least once every 6 months should be maintained to assure good cell electrode life. This low cleaning frequency is often considered an advantage of the tubular cell design.

5.4.1.2   PEPCON — ChlorMaster™

Pacific Engineering and Construction (Pepcon) developed an ammonium perchlorate manufacturing electrolytic cell, the product used for solid rocket fuel. This cell utilizes a carbon electrode inserted into a titanium tube.

The cell is a monopolar design with only one large diameter lead dioxide coated anode and one closed end titanium tube cathode. Although lead dioxide is not as electrochemically efficient as DSA related coatings, the coating was effective in producing sodium hypochlorite from seawater. Given the lack of conversion efficiency of this coating, cell power consumption is approximately 25 to 50% greater than competitive cells operating under the same conditions. The loss of the lead dioxide during operation of this cell did not cause a cell malfunction because carbon will perform well as an anode. Carbon will, however, gradually dissolve, increasing the cell operating voltage and therefore increasing operating power consumption.

Cathode deposits are more rapid in this design compared to the other designs due to the lower seawater flow velocity. The cell flow inlet and outlet are in the top of the cathode tube causing very low circulation conditions. Cell modules are usually designed with vertically mounted cells series both hydraulically and electrically. Multiple cell racks are combined for complete systems to meet various disinfection demand requirements. Pepcon recommends that deposits be removed using a 5% nitric acid solution.

Pepcon recently designed and built a cell using DSA type coated titanium plate material in a pipe arrangement. This cell uses multiple flow paths over the electrodes to reduce cathodic fouling. The cells are mounted horizontally on a rack one above the other where a feed header will distribute the seawater into the bottom cell at multiple points for flow over the electrodes. Flow exits from the top of the horizontally mounted body into the cells above in series and discharges the sodium hypochlorite solution and gas into a product header.

Few plants employing this cell are known by the authors to have been installed. Plate cell testing began in the early 1970s as a spin-off technology of chlorine cell dimensionally stable anode technology by Diamond Shamrock, DeNora of Italy, and Mitsubishi of Japan. Each of these companies was affiliated with the DSA business for chlorine cell electrodes.

5.4.1.3 Diamond Shamrock — Sanilec®

A patent was awarded to Diamond Shamrock in 1975, U.S. Patent 3,893,902, for a plate cell configuration. This cell is a mono-polar design using DSA coated expanded titanium anodes and titanium or nickel alloy Hastelloy® C plate cathodes. The flanged cell body is either machined PVC or molded PP having titanium internal spacers and fasteners while using silicone and Viton® rubber o-ring seals. Electrode spacing is controlled by swage PVDF buttons in the expanded titanium anodes. The cell cover is unique. It is acrylic material allowing the user, when mounted vertically, to see the cell operation or deposits forming on the cathodes.

These cells can be mounted vertically on racks with interconnecting piping and bus bars or mounted horizontally with the cover face-to-face stacked flange-to-flange four cells in height. Most installations are in the rack-mounted configuration in many cooling water applications. Offshore installations are nearly all horizontally mounted as a means of space conservation. The horizontal arrangement also has the advantage of reducing interconnecting cell piping.

Cell cleaning is necessary once a month unless the system design is modified to accommodate higher cell seawater flow velocities. Because the system is designed with an acid system as part of the equipment supplied and the cleaning duration short, 4 h of cell cleaning is not usually considered a maintenance problem.

Diamond Shamrock began cell testing in Peabody, Massachusetts where the cell was installed with expanded metal anodes and cathodes. After a few days operation it was apparent that the expanded metal cathodes had to be replaced because cathode deposits built up extremely quickly.

Another major discovery at the Peabody site was the effect of manganese on anode performance for hypochlorite production. Manganese as a dissolved solid in the seawater reacts on the anode surface to form manganese dioxide. A manganese dioxide contaminated surface reacted electrochemically to form oxygen rather than chlorine, resulting in a loss of cell efficiency. Acid cleaning for the removal of manganese dioxide is difficult. Therefore, a test was designed to reverse the polarity of the anode to remove the MnO_2 deposit. Note the design was not for reverse current. To reverse the current would dissolve the cathode into granular material in the cell.

As a note of information, cell assemblies installed with the electrode current reversed will, with alloy cathodes such as Hastelloy® C, dissolve the cathodes in less than 8 h of operation. Titanium is not as susceptible to current reversal cathode failure because the necessary polarity voltage is approximately 3 times the voltage for accepted alloy cathodes. Also the rectifier normally will not have the voltage capacity to pass current when cell cathodes are titanium.

5.4.1.4   DeNora — SEACELL®

The SEACELL was developed by Oronzio DeNora Spa of Milan, Italy during the 1970–1974 time period. This cell is a bipolar plate electrode assembly within a tubular filament wound FRP body having a PVC liner. The electrode coatings are DSA because DeNora was part of the original group of companies responsible for DSA coating development patented by independent researcher Henry Beer of Belgium. Electrode spacers are PVDF and the cell heads are rubber-lined steel.

Power consumption ranges from 3.5 to 4 kWh per kilogram. This power consumption is achieved by increasing the anode surface area to reduce the current density and thus, the cell voltage. Lower current density in turn reduces the power consumption associated with bipolar titanium cathodes.

Small capacity cells are mounted horizontally in multiple units to meet disinfection requirements. Large capacity cells are singly mounted vertically. Generally, DeNora has focused on the large capacity electrolytic systems.

5.4.1.5   Mitsubishi Heavy Industries — Marine Growth Preventing System

Mitsubishi Heavy Industries (MHI) became an electrolytic cell manufacturer as a result of building large power stations. The MHI cells utilize a rubber-lined steel casing and cover, platinum coated titanium anodes, bipolar electrodes and Hastelloy C terminal cathodes. The electrode gap is 0.118 in. (3 mm). Power consumption is usually greater than 4 kWh per kg of product at the usually accepted electrode current density of 1 amp/in² (1.55 kA/m²). Seawater flow through the cell is 123 gpm (28 m³/h) and the final sodium hypochlorite concentration produced by the cell is approximately 1500 mg/l (0.15%) at the rated operating current.

Electrode spacing and the use of electrode modifications to change the cathode current density at the edges reduces the acid cleaning requirement for this cell design to once every 6 months. U.S. Patent 3,645,880 claims cathode deposit reductions because the cathode gaps are 5 times the electrode gaps at each end of the anode. The cells are mounted in numbers up to 12 and connected in series hydraulically and electrically.

5.4.1.6 Daiki Engineering — Hychlorinator

Daiki cells are designed as bipolar cells contained in a rubber lined steel body. Covers are gasketed PVC. The electrodes consist of platinum coated terminal electrodes, platinum coated bipolar electrodes, and a Hastelloy C terminal cathode. Coating life is suggested to be 5 years for Daiki's platinum coating.

The electrode gap is quite wide at 0.14 in. (3.5 mm). Power consumption is 1.8 to 2 kWh/lb of chlorine produced. This power consumption indicates the current density is below that of competitive cell designs. Reduced current is necessary to achieve the low power consumption figure, discussed above, as a result of platinum coating's lower chlorine conversion efficiency and the wide electrode gap.

Seawater flow through the cell is 186 gpm (42 m^3/h). This flow rate is relatively high. Chlorine concentrations in the area of 1000 mg/l (0.1%) are normal for cells operating at these flow rates. According to the literature acid cleaning is not required.

5.4.2 Cell module groupings

Electrolytic cell modules consist of a group of cells connected together hydraulically and electrically in series to form a complete cell circuit. The cells in these groupings can be oriented in the vertical or horizontal position. From an electrochemical perspective, a cell will provide product regardless of its orientation as long as proper operating conditions exist. Cell cathode deposit formation is a function of seawater velocity, specific seawater conditions, operating current, and hydrogen level; therefore these are the primary cell circuit design considerations.

5.4.2.1 Vertical cell circuits — seawater electrolysis

The combination of cell design, cell flow arrangement, and hydrogen production make vertical cells a desirable system configuration for all cell manufacturers supplying seawater systems. Interconnecting piping and electrical bus bars are brought together in this configuration primarily to minimize hydrogen pocketing in the piping system and bus bar costs.

All manufacturers use variations of the vertical cell arrangement in their designs. How the cells are configured in a circuit is dependent on individual cell output.

Vertical cell configurations assure that produced hydrogen always enters the cell bottom and exits the cell top. In the case of a low cell flow condition, the cell will be flooded should the water flow sensor not stop system operation. This arrangement requires additional interconnecting piping. Electrical bus bar needs are not reduced in this configuration because the power requirements are the same regardless of cell orientation.

Hydrogen is normally passed through the cell circuit. Early Diamond Shamrock Sanilec cell modules contained a hydrocyclone to remove the hydrogen interstitially from the cell circuit. There is a small power consumption savings (1–3%) when hydrogen is removed in this manner. The removed hydrogen is then inserted into the product discharge line to be carried to a shock dosing tank for air dilution, through a second removal cyclone for discharge to atmosphere or sent directly to the dosing point. The authors are not aware of other manufacturers practicing interstitial hydrogen removal in vertical cell circuits. This practice was employed in early Sanilec designs to improve large project power consumption for very large electrolytic systems.

Frame assemblies are painted structural steel or fiberglass reinforced plastic (FRP). In all cases the desire for corrosion control is an important client consideration whether indoors or in direct weather conditions. Three and four part paint specifications suitable for offshore environments should be considered regardless of site conditions.

Because FRP frames are not prone to corrosion one should only be concerned with the assembly fasteners as noted below. Careful inspection of the support members is also essential. FRPs structural properties require added support members to assure good frame stability. System suppliers have provided painted and unpainted FRP frames. Painted frames are given a light abrasive blast surface roughening and then coated with a suitable epoxy finish.

As mentioned above, fasteners are an important issue. This issue is not readily apparent for a new system. Yet as maintenance is required when the installation ages, the importance of high quality fastener materials becomes readily apparent. The authors recommend using 316 stainless steel fasteners for frames, piping, and associated equipment. Using 316 stainless steel will allow for less system maintenance in the future. If coated mild steel fasteners are used in systems installed in open weather, they will corrode and make disassembly difficult. Steel anchor bolts in concrete will eventually cause the concrete to crack as the rust expands the original bolt size. This recommendation applies to all equipment exposed to seawater and sodium hypochlorite.

5.4.2.2 Horizontal cell circuits — seawater electrolysis

The same equipment and conditions are used to design circuits having a horizontal cell orientation. It is the practice of Electrocatalytic (ELCAT) to install multiple tubular cells horizontally in multiple horizontal circuits. This configuration allows the designer to provide a single large rectifier serving many cell circuits. Horizontal tubular cells are easily assembled on racks one above the other, having a common inlet seawater supply header and product discharge header.

No hydrogen removal within the circuits is necessary for the ELCAT cells in the horizontal configuration. The hydraulic velocity and number of cells in series eliminates hydrogen removal.

Electrical design configures each parallel cell circuit with the negative DC connections on the inlet and outlet cells. Positive DC power is supplied to the center of the circuit and split in each direction to the inlet and outlet cells. A monitoring system is necessary to assure the cells are operating with relatively balanced current on each parallel circuit. Should one or several circuits electrically in parallel have excessive resistance, the low resistance circuits will operate at high current. High current will then cause shortened anode life and much higher cathode deposit rates.

DeNora utilizes horizontal configurations when employing smaller cells in series. Very large cell systems utilize a vertical cell orientation. These cells are extremely large requiring large overhead cranes for cell removal and maintenance.

DeNora's cell design employs vertically oriented plates in a horizontal tube where the hydrogen escapes vertically to the headspace above the plates, travels through the cell to the following cell in the circuit, is removed after the second cell of the circuit where there are more than two cells in the arrangement, or is passed in the cell stream to the gas disengaging/dosing tank.

Hydrogen is removed interstitially in DeNora's medium-sized cell module configurations. The hydrogen is re-inserted into the product discharge line to be carried to a gas disengaging/dosing tank for air dilution, or sent directly to the dosing point. The authors are unaware of any other manufacturers practicing interstitial hydrogen removal in horizontal cell circuits.

5.4.2.3 Horizontal cell circuits — brine electrolysis

All brine electrolysis cells presently available have a horizontal orientation. Each cell, regardless of manufacturer, has several compartments in each body. A single cell or several cell assemblies are connected in series hydraulically and electrically. The cell is designed to assure that the hydrogen separates to the top of each compartment and flows through the cell to discharge with the product to the next cell or to product storage.

Electrode arrangements vary by being oriented either vertically or horizontally, with vertical the preferred arrangement. Horizontal arrangements require the electrodes to be expanded metal to assure effective hydrogen release from the electrode assembly. These cells perform well as long as the system is properly maintained with a strict cell cleaning maintenance schedule. Vertically oriented electrodes are generally made of sheet material since the hydrogen will release quickly due to the hydrogen/water specific gravity difference. Sheet material also provides easy bipolar electrodes where one face or one end is anodic and the opposite is cathodic.

Recently patented developments by Severn Trent Services' ClorTec division have revolutionized brine electrolyzer performance through the use of modified cell flow schemes, feed water cooling, improved performance anode coatings and cathode coatings. The combination of these changes has increased efficiency from 65% to over 80% current efficiency, reduction of DC power consumption to 1.8 DC kWh per pound of chlorine, reduction of by-product chlorates from 400 mg/l to less than 150 mg/l and, most importantly, reduced salt consumption from 3.5 to 3.0 lb of salt per pound of chlorine produced.

5.5 Electrolysis systems

5.5.1 Equipment

As outlined in the brine and seawater system general descriptions above, each part of the electrolysis system has an important role in the sodium hypochlorite production. All of the equipment listed is essential to assure a complete and operationally reliable package.

The need for the electrolyzer cell and DC rectifier is obvious since, without both pieces of equipment, no electrolysis is possible. The following narrative is a discussion of the ancillary equipment to the cell/rectifier combination.

It is not essential for brine systems to include a water softener and have high quality salt for the cell-rectifier system to function. However, if these recommendations are not followed, the cathodic deposits from the salt and water may accumulate in the cell. These deposits will close the electrode gap, causing reduced electrolysis and electrode degradation. The same problem will occur in seawater cells if the seawater flow rate is not maintained at the supplier recommended level. Consequently, frequent acid cleaning is necessary to protect the electrodes. Otherwise, a probable reduction in cell electrode life could result. Such deposits retain an alkaline layer on the anode surface that renders the coating inactive. This coating activity loss is referred to as coating passivation.

Cell cleaning is an essential part of a system maintenance program. Salt, water, and seawater contain calcarious material that reacts with the cathodic alkaline layer. Calcium and magnesium are the elemental materials that cause the deposits. The predominant deposit in brine systems is calcium carbonate, while the seawater system's deposit is primarily magnesium hydroxide. These different deposits are caused by differences in the cell feed brine. Seawater has high magnesium levels and an average pH of 8.5. Brine system feed has high calcium levels and essentially a somewhat neutral pH since drinking water is the source water. Although both deposit forms are easily removed by hydrochloric acid, the carbonate form in the brine system is more quickly dissolved, shortening the cleaning cycle. Most electrolysis systems are supplied with an acid cleaning system consisting of a tank, pump, and associated piping and valves. Note that cell cleaning must be performed on the schedule recommended by the electrochlorinator vendor to assure an adequate life from the electrolyzer cell system.

Hydrogen dilution fans are considered part of all systems, yet their importance is defined by the system location and local conditions. For example, many systems are designed without fans because they are located away from ignition sources. In some cases, the hydrogen is vented from the degas device directly to the atmosphere through a water seal. Other systems use open top tanks to allow the hydrogen to separate and disperse. Brine systems typically have closed tanks; therefore fans are required for normal dilution, although some tanks may be directly vented.

Although pumps typically require very little maintenance, they are essential to effective system performance. Whether they supply water to the system or product to the dosing point, routine maintenance is required to assure system reliability.

Brine systems will normally have only product dosing pumps. Generally, they are positive displacement pumps to overcome system dosing point pressures. Variable speed drive controls are common to assure correct product addition. Some systems may require a centrifugal pump on the inlet water supply to boost the pressure due to low resident system pressure. These pumps interface with the control scheme for operation whenever the system is in operation.

Seawater systems will normally have centrifugal pumps on the inlet water supply to meet flow and pressure requirements in the system. Systems having low dosing pressure and no shock requirement seldom have dosing pumps. A substantial number of systems require centrifugal pumps to meet continuous and shock dose chlorination requirements. Tanks of sufficient size are provided for hydrogen removal and for a positive pump suction head.

5.5.2 Instruments

System instruments are designed to provide continuous system monitoring for conditions outside acceptable operating limits. They also form an essential part of the control and safety logic.

All electrolysis systems are designed to have water and brine flow indication, cell level sensors, cell temperature sensors, tank level indication, hydrogen fan indication, and dosing point residual indication. All of these sensors can, depending in the vendor's control logic design, have alarm and shutdown set points. Systems from all vendors will have the same general alarm and status indicators to confirm proper operation. With few exceptions, each of these alarm situations will shut down system operations to prevent serious damage to the equipment or hazard to the personnel. The following discussion will provide general guidelines for specific alarm conditions.

Cell operation is prevented during filling at start up and after acid cleaning. This lockout mode assures that the electrodes are always covered by brine solution. Should the electrodes become exposed while generating hypochlorite, hydrogen detonation may occur. Although the electrodes may be partially immersed in brine solution, a spark potential may still exist. Such actions as closing the cell discharge valve should immediately stop cell electrolysis if the flow switch is functioning properly. If not, the cell liquid level will quickly be depressed exposing the electrodes to hydrogen by the generated gas pressure. In addition, the cell will have some increase in oxygen production due to the brine depletion from loss of flow. Cell level switches normally are not used on seawater systems. Therefore, the flow switch is even more important in seawater systems than in brine systems.

Water and brine flow meters are essential to confirm a generally accepted water to brine ratio of 10:1. Flow alarms are required to be certain that the cell temperature does not exceed 140°F (60°C) or cell damage will occur. The addition of a temperature sensor is part of a comprehensive safety system to stop cell electrolysis should the flow sensor fail.

Tank level instruments are designed to operate the system when refilling the tank is necessary, protect the dosing pumps, and prevent tank overflow. Generally, two operating set points and two alarm points are used for tank level instruments. The operating set points maintain the tank level by operating the system via the control scheme. The low-level alarm set point will stop the pumps and alarm while maintaining generator operation. A high-level alarm setting will alarm and stop generator operation.

Hydrogen dilution fan switches will alarm and stop generator production. The dilution fan switches assure that the large volume of hydrogen in tanks is always properly diluted. Systems without fans are also without air flow switches. The location of open air system hydrogen venting is important to assure that safe atmospheric dilution occurs.

Residual analysis or oxidation reduction potential (ORP) analysis is often applied to control dosing pumps or, in the case of seawater systems, DC rectifier operating current. Analog signals are used through a PID controller to assure stable operation of the pump or rectifier. Historically, residual analysis has been the control method of choice, however, as instrumentation becomes more sophisticated ORP is becoming more accepted. The use of chloramines for residual maintenance in water systems requires the comparison of free to combined chlorine concentrations.

5.6   Materials of construction

5.6.1   Plastic materials

5.6.1.1   PVC (polyvinyl chloride) and CPVC (chlorinated polyvinyl chloride)

These two materials are suitable for all applications within the realm of on-site hypochlorite generation, system supply potable water, seawater, and concentrated brine (300 g/l strength or 26.39% salt by weight, brine basis) and sodium hypochlorite product up to 15 weight percent (150 g/l). The maximum recommended upper operating temperature limit for this material is 140°F (60°C).

Although all U.S. suppliers of these materials combine ultraviolet (UV) inhibitors in their products, care must be taken in some extreme sunlight environments to prevent UV degradation of PVC or CPVC. Ultraviolet degradation exhibits itself via a white surface discoloration, loss of material elasticity, and a reduction of impact resistance. Ultraviolet degradation is prevented using a white latex paint coating. Pipe surface preparation is provided using materials such as methyethyl keytone, methanol or acetone.

Joining methods are important to consider with these materials. The most common joining methods are solvent cementing, threading, and flanging. Natural stress relieving properties inherent in PVC and related polymer compounds, with few exceptions, generally prevent reliable threaded joints.

Threaded connections present the most difficulty in achieving reliable leak-free performance. Field experience has proven that commonly used Teflon tape causes the threaded female fitting to randomly split and leak. Several commercially available joining compounds have also been used with sporadic results. The compound found most reliable for PVC threaded joints is manufactured by Rectorseal of Houston, Texas under the market trade name of Rectorseal 5. This material is a soft set compound acceptable for a broad spectrum of pipe materials, is UL classified, suitable for a variety of service including potable water, and is non-toxic in accordance with NSF 61 standards.

Solvent cementing the joints is the most reliable sealing method. While sloppy joining techniques may be used for water, seawater, and salt brine solution, supplier recommended joining solvents and cements, joint preparation methods, and cement application methods must be adhered to strictly to assure long-term reliability in sodium hypochlorite service. Sodium hypochlorite strengths of 0.5 to 1.3% generated from dissolved salt brine may degrade the cement causing eventual seepage in poorly prepared and assembled joints.

The use of CPVC is generally regarded as an unnecessary additional cost compared to PVC. CPVC will average two times the price of PVC for pipe, valves, and fittings with no apparent service advantage over PVC in on-site hypochlorite applications. In instances where CPVC material was utilized, frequent failures have occurred during shipping and field storage. These failures resulted in delayed commissioning on overseas projects. Analysis indicated the failures were the result of stress breakage due to the harder yet more brittle CPVC material in areas of true union valve connection nuts and flange cement joint sockets. While there have been similar experiences with PVC, they were rare.

5.6.1.2 *Polypropylene (PP)*

This material is suitable for more limited applications within on-site sodium hypochlorite generation operations, system supply water, seawater, and concentrated brine (300 g/l strength or 26.39% salt by weight, brine basis).

Polypropylene is **not recommended** by the authors for sodium hypochlorite service. Experience has proven that sodium hypochlorite may adversely affect the long-term reliability of polypropylene. The upper operating temperature limit for this material is 180°F (82°C).

Long-term operation of welded parts in the adjacent weld effect area and machined parts in the region of sharp corners such as on pump housings are very prone to radial and stress corrosion cracking. Although parts that have been molded do not have the same propensity for this condition, caution should be exercised for all sodium hypochlorite applications. Some U.S. suppliers of these materials combine an ultraviolet (UV) inhibitor in their products by applying black pigment in their compounding.

Because this thermoplastic cannot be dissolved by the strongest solvents, the most common joining methods are thermosealing, threading, and flanges. The natural stress-relieving properties inherent in polypropylene, with few exceptions, generally prevent reliable threaded connections. Fusion welding and hot gas welding are the preferred methods for socket type joints. Careful application of the manufacturer's procedures is essential to assure proper bonding and pressure ratings.

5.6.1.3 Acrylonitrile butadiene styrene (ABS)

This material is suitable for water, brine, and raw seawater applications within the realm of on-site sodium hypochlorite generation. It is also suitable for system supply water, seawater, and concentrated brine (300 g /l strength or 26.39% salt by weight, brine basis) exposure. Manufacturers are not willing to recommend the use of this material for sodium hypochlorite service due to long-term service failure prospects. This material's upper operating temperature limit is slightly higher than PVC at 160°F (71°C).

Joining methods are important to consider with this material. The natural stress-relieving properties inherent in polymer compounds, with few exceptions, prevent reliable threaded connections. The most common joining methods are solvent cementing, threading, and flanging.

5.6.1.4 Polyvinylidene fluoride (PVDF, Kynar®) and polytetrafluoroethylene (Teflon®)

These materials are suitable for all applications within on-site hypochlorite generation systems, e.g., system supply water, seawater, and concentrated brine (300 g/l strength or 26.39% salt by weight, brine basis) and sodium hypochlorite product up to 15 weight percent (150 g/l). The maximum recommended operating temperature for these materials is 280°F (138°C).

Joining methods are limited for these materials. The natural stress relieving properties inherent in polymer compounds, with few exceptions, prevent reliable threaded connections. Of these materials, Teflon is a particular offender due to cold flow. Because the strongest solvents will not dissolve this polymer, the most common joining methods are thermosealing, threading, and flanges.

5.6.1.5 FRP (fiberglass reinforced plastic)

This material is suitable for all applications within on-site hypochlorite generation systems provided the correct resin materials, recommended by manufacturers, are used to assure excellent life characteristics. The resin of choice, regardless of the manufacturer, is a vinyl ester based with a BPO-DMA (benzoyl peroxide-dimethylaniline) cure catalyst. The upper operating temperature limit is 140°F (60°C).

Although all U.S. suppliers of these materials combine ultraviolet (UV) inhibitors in their products, care must be taken in extreme sunlight environments to prevent UV degradation. Ultraviolet degradation exhibits itself via surface resin degradation exposing the outer resin layer of chopped fiberglass. This degradation can be prevented by painting the pipe with white latex paint. The pipe surface preparation is provided with such materials as commercial solvents or slight surface roughening with sandpaper.

Joining methods are very important for this material. The natural stress-relieving properties inherent in other polymer compounds will seldom be exhibited for FRP threaded connecting joints. The most common joining methods are epoxy resin cementing, threading, and flanging.

Epoxy resin cement joints are the most reliable; however, sloppy joint preparation and epoxy application will cause seep leaks through the joint resin. Therefore, one must strictly adhere to the supplier recommended joining epoxy materials, joint preparation methods, and epoxy resin application techniques to assure long-term sodium hypochlorite system service reliability. Experience has proven that 0.5 to 1.3% sodium hypochlorite, generated from dissolved salt brine, will attack poorly assembled joints causing eventual failure.

5.6.2 Elastomeric sealing materials

5.6.2.1 Fluorocarbon elastomer (Viton®)

This fluoroelastomeric material is a copolymer of hexafluoropropylene and 1,1-difluoroethylene. It is suitable for all application areas of sodium hypochlorite systems. The temperature limits are −13°F (−26°C) to 350°F (176°C).

5.6.2.2 Ethylene propylene diene methylene (EPDM)

This material is an ethylene-propylene diene methylene compound blend elastomer suitable for application in system feed water, seawater, all strengths of sodium salt solutions and very low strength hypochlorite solution. The temperature limits are −65°F (−53°C) to 300°F (150°C). Material usage includes valve o-rings, valve seats, and full-faced flange gaskets. This material is not recommended for use in hypochlorite service. Only full-faced flange gaskets provide acceptable long-term service in sodium hypochlorite systems.

5.6.2.3 Buna-N

This material is a copolymer of acrylonitrile and butadiene suitable for application in system feed water, seawater, and all strengths of sodium salt solutions. The temperature limits are −120°F (−50°C) to 250°F (120°C). Material usage includes valve o-rings, valve seats, and full-faced flange gaskets. Once again use only the full-faced flange gaskets in hypochlorite service for reliable service.

5.6.2.4 Neoprene

This material is a synthetic rubber in use for over 70 years as suitable for application in system feed water, seawater, and all strengths of salt solutions. The temperature limits are −10°F (−23°C) to 200°F (93°C). Material usage includes valve o-rings, valve seats, and full-faced flange gaskets only as noted for the above application services. Use this material only for full-faced flange gaskets in hypochlorite for reliable service.

5.6.3 Metals

5.6.3.1 Titanium (Ti)

This metal is impervious to all materials used to generate sodium hypochlorite product. Titanium is the substrate for the DSA anodes (positively charged electrode) and, in most cell configurations, the cathodes (negatively charged electrode). Titanium may be considered to be electrolysis cell exposure tolerant.

Titanium of varying purity can be used in most sodium hypochlorite applications. However, manufacturers of sodium hypochlorite equipment generally use chemically pure (CP) grade 2 or better material because specialized metallurgical properties are not required.

Welding titanium should be carried out under an argon gas atmosphere using a TIG (tungsten inert gas) torch with parent metal or wire filler metal. MIG (metal inert gas) welding using a wire feed system can cause feed reliability problems. Titanium has a surface texture resulting from the wire extrusion process causing wire guide abrasion and subsequent wire feed failures. Good back shielding is essential to assure contamination-free welded joints having parent metal ductility.

Material cost and specific needs define the best application for this material (e.g., cell electrodes and fasteners). Although some electrochlorination facilities have been installed using titanium centrifugal pumps, titanium is seldom used beyond cell internal component applications.

5.6.3.2 Hastelloy® C 276 (Hast. C)

This metal is also impervious to all materials used to generate sodium hypochlorite product. This alloy will tolerate the severe corrosion environment of hypochlorite applications.

Hastelloy is available in varying alloys. Manufacturers of sodium hypochlorite equipment generally recommend Hastelloy C 276 grade since its special properties offer excellent service life.

Welding Hastelloy C should be carried out under an argon gas atmosphere using a TIG (tungsten inert gas) torch with parent metal or wire filler.

MIG (metal inert gas) welding is also used with specific wire alloys to assure proper weldment ductility. As with titanium, good back shielding is essential to assure contamination free welds.

Material cost, electrical properties, and corrosion resistance make this material excellent for seawater cell cathodes and pumps. The seawater electrolyzer cell is virtually the exclusive domain where Hastelloy C 276 is used as cathode material due to high cost when compared with less expensive alternatives.

5.6.3.3 316L stainless steel (316L)

This metal is only used on the seawater supply portion of seawater electrolyzer systems. Material cost, corrosion resistance, and specific applications define this material's use as seawater booster pumps and seawater strainers. Many electrochlorination facilities have installed horizontal or submersible centrifugal pumps and manual or automatic strainer assemblies using this material for acceptable service life.

5.7 DC power rectifiers

Three major types of rectifier units used in electrocatalytic sodium hypochlorite generation are described in this section. Each of these rectifiers has its own merits based upon needs specific to operational and environmental requirements.

Tap switch voltage control rectifiers — These units have no accurate current control mechanism. Rectifier control is provided from transformer primary tap points using a multipoint click switch regulator. The regulator varies the input voltage to the transformer resulting in a secondary voltage variation. For this reason, the rectifier current varies as the solution temperature or salt content changes the cell resistance. For units of this type it is very important that the system has a closely maintained feed water temperature and salinity level.

Thyristor rectifiers — Thyristor rectifiers, often called SCR (silicon control rectifier) controlled, are the most commonly used for electrolysis cells where more than 100 amp DC service is required. Thyristor designs allow the use of AC input voltages up to 11,000 V and normal 50 or 60 Hz requirements. Most thyristor rectifier designs operate in the range of 75 to 90% efficiency at a power factor of 0.7 to 0.85 when at full load.

Switching power supplies — Switching power supplies operate at a high frequencies, normally in the range of 10 to 200 kHz. While there are large units that will accept 480 V 3 phase input power, normally this type of rectification operates with 110/220 V AC single phase power. High frequency devices will have a power factor of 0.99 and efficiencies of 75 to 85%. Higher output systems will eventually be available and should be utilized to assure further reductions in operating cost. Switching power supplies offer the advantage of a very small transformer resulting from a high operating frequency.

5.7.1 Rectifier metering, control, and operation

Switching power supplies manufactured in the United States normally are supplied with digital meters because the system is fully electronic. Analog meters are generally used for tap switch and thyristor type DC rectifiers. When digital meters are used for tap switch and thyristor rectifiers, additional electronic control cards are essential for proper meter function.

Tap switch rectifiers have very simple controls. The current is adjusted via individual AC line taps. These line taps adjust the transformer secondary line voltage by making the transformer slightly out of balance between each phase. The current requirements for proper cell operation are based upon the cell resistance. The resistance is controlled by the feed water salt concentration, feed water temperature, and the nominal cell electrode resistance. Any variation in these parameters will affect the cell operating current, increasing current for reduced cell resistance and decreasing current for increased cell resistance, which is caused by a decrease or increase in cell voltage.

SCR power supplies have numerous configurations. The two dominant configurations are for rectifiers having thyristors (SCRs) on the AC line before the transformer and those having thyristors (SCRs) after the transformer. Both types of units are controlled via electronic circuit boards having internal safety circuits and both DC voltage or current control capability. These devices are designed to have 6, 12 or, rarely, 24 pulse DC outputs. The purpose of 12 and 24 pulse units is to improve power factor and to reduce harmonic distortion on the AC power line. Designers accomplish 12 and 24 pulse outputs by phase shifting the secondary output using Delta-Wye and Delta-Delta configurations for the main power transformers.

Switching power supplies are all electronic by design, thus, have similar controls to SCR type supplies. Most units are designed for DC voltage control that relates to their usage in the computer power supply industry. Units supplied for on-site sodium hypochlorite electrolyzers require constant current with floating voltage. Therefore, manufacturer modifications are usually made to the power supply for a current controlled design.

5.7.1.1 Cooling

While a myriad of cooling schemes exist in sodium hypochlorite generation (e.g., air, water, and oil), by far the most common cooling method is fan operated forced air cooling. Local environmental and safety conditions will determine the cooling method most acceptable for reliable equipment life.

Rectifier units are designed to operate at maximum ambient temperatures of 105°F to 125°F (40°C to 50°C) to assure that control and power electronics can survive possible installed environmental service stresses. All DC power systems are designed around these temperatures, unless more stringent specifications are necessary resulting from local conditions.

DC power equipment installed under cover as partial or full environmental coverage can be designed to rather low National Electrical Manufacturers Association (NEMA) classifications. These types are most often used in plant applications where local weather conditions are the primary protection consideration. Forced air is the preferred cooling method with equipment having one or more high volume low-pressure fans exhausting from the top of the power component cabinet. Cabinets are built to NEMA 1 or NEMA 3 (IP 23 to IP 32) cabinet classifications for air-cooled applications.

Side exhaust units are built to allow uncovered outdoor installations, however, the manufacturer must increase fan size approximately 25% for efficiency losses due to the air flow direction change required. Also, the side mounted cover design must minimize additional air pressure losses.

Rectifier DC device and transformer cooling efficiency is dependent upon the internal cabinet equipment location. DC devices require 500 to 1000 linear ft/min air passage over the cooling heat sinks. Baffling arrangements are usually installed to direct the flow of air over the thyristor heat sink.

Where the transformer is mounted inside a single enclosure below the DC devices, manufacturers design for transformer heat losses. Present transformer materials for construction such as Nomex are very temperature tolerant. Transformers designed and manufactured using these materials provide excellent life cycle performance with minimum air flow requirements.

Isolated instances have occurred where systems were designed for much higher temperatures than discussed above. Installations in the Middle East, Saudi Arabia, for example, normally require a sun-shade and independent cooling in the control cabinet to prevent premature control electronics failure. Thermometers placed in control cabinets in full sun have demonstrated temperatures above 175°F (80°C). Unless electronic components are mil-spec equipment, early failure is assured.

Cooling methods for indoor or outdoor temperate locations are normally forced air having the air exit from the top or side of the thyristor power cabinet. Screening is placed in the bottom of the cabinet. Course filters also are often added to the cabinet bottom for efficient debris ingress control in the air stream.

DC power equipment in severe environment installations for seawater electroclorination requires special attention. The same is true for offshore oil applications, oilrig and floating production storage operation (FPSO) cooling loop, and offshore water flooding. These are plant applications where the designer must account for severe temperatures and seawater mist-laden atmospheres.

Designers must consider the use of sealed circulating forced air cooling where an external water source, seawater or potable water, is applied to an internal water/air exchanger for controlling the internal Btu loads released by the power components. Designs include units using oil cooling in an external transformer remote from the DC power devices and a closed-loop water to air circulation systems for cooling thyristor DC power devices. When the power requirements are not excessive, the system is designed with external isolated thyristors mounted to a cooling block that can accept virtually any water cooling source.

Oil immersion cooling is used on rectifiers where the environment is severely corrosive. Desert and offshore locations are prime examples of temperature, airborne dust and sand, and salt laden atmospheric conditions where oil immersion prevents premature equipment failure. The design principles applied are identical large power transformers. The internal components are rack mounted for easier removal and immersed from the top in a tank of oil with external cooling coils to provide convection circulation of the oil. Note that the term "easier" means that one must bring a portable crane to the site to lift the rack from the oil tub rather than draining the oil from the tank and removing side panels. When designing cooling systems, the authors recommend that the engineer ensure 1) there is no need for air fans and 2) the manufacturer has provided sufficient convection cooling surface area for the oil cooling coils. Thyristor devices are mounted in the rack arrangement to assure exposure to the circulating oil after it is cooled in the exchanger. All the equipment is designed to 50ºC or 60ºC ambient temperatures as a further reliability factor.

5.7.1.2 Area classification

Electrochlorination systems designed for applications in refineries and offshore platforms require particular attention to temporary or constant explosive gas environmental conditions.

Control equipment for area classifications NEMA class 1 Division 2 Groups C and D are most often configured with a Z purge wherein the controls are always under instrument air pressure of several inches water column. Normal air requirements are 1 to 3 ft^3/h to maintain the pressure protection with a sealed cabinet. Several manufacturers have UL approved air control instrumentation to meet mandated flow and pressure specifications while providing a system safety shut down should pressure loss occur.

NEMA Class 1 Division 1 Groups A and B areas where explosive conditions can exist continuously do not normally allow Z purging. These conditions require NEMA 7 sealed aluminum control enclosures, frequently referred to as coffins, because they are aluminum castings designed to contain an explosion within the enclosure while sealed.

DC power components do not require stringent control enclosure or purging requirements noted above because DC components are non-arcing devices when compared to relays and control contactors. Cabinetry used for these applications is sealed only to minimize the ingress of water and debris.

5.7.1.3 Maintenance

While minimal maintenance is required for water and oil cooled systems, regular oil and closed loop cooling media inspections must be performed. Each manufacturer should provide inspection guidelines to assure that the inspector will detect problems before they become equipment failure incidents. Forced air cooled systems require quarterly cooling fan and motor inspection, quarterly filter screen cleaning, semi-annual wire and cable termination connection inspection, and DC device and transformer cleaning. It is best to use a compressed air nozzle for cleaning transformers and DC devices.

5.7.2 DC rectifier operating status

The DC rectifier is a singular stand-alone unit connected to an electrolyzer cell circuit in sodium hypochlorite generation systems. While these units may be operated from remotely located control panels, each unit should have its own independent controls and safety equipment.

Rectifier unit controls and DC output current and voltage meters indicating operating levels are on the face of the rectifier panel. Internal temperature sensors for the thyristors, diodes, and transformer are interlocked with the AC power supply controls to provide unit shut down if there is excessive temperature rise. Alternating current safety circuits meeting NEMA motor control safety requirements are normally supplied in the rectifier system, eliminating the need for remote motor control connections unless required by local regulations.

Large rectifier units are designed with DC fusing to prevent a direct AC phase-to-phase short fault caused by a DC device failure. These are high speed fuses installed in line with each device and will often have an indicator fuse across the circuit to indicate the failed fuse and shut down input power until the problem is corrected.

Remote DC output operating status is easily indicated on an operator interface panel via analog signals from rectifier internal electronic interface cards. Interface electronics are also available to provide DC current control from the PLC. System voltage must float, thus voltage control is not required. These arrangements are normally add-on cards allowing the use of standard rectifier control electronics. Care must be used when controlling from the PLC panel with feedback control systems. Because the system may oscillate, a PID control loop must be added to assure control stability. While the control loop can be part of the PLC program, one should consider the use of a small PID loop control unit. This type of unit allows one to program sensitivity changes in the control loop without the need for PLC program changes.

5.7.2.1 DC rectifier independent alarm conditions

Alarm conditions that cause shut down and alarm are high SCR temperature, high diode temperature, high transformer temperature, high transformer oil temperature, SCR fuse failure, diode fuse failure, fan motor fuse failure, and low fan pressure, where applicable.

5.7.3 DC rectifier safety equipment

Direct current power supply units are normally supplied as stand-alone equipment with all required operational and safety equipment included as part of the package. Cabinet design standards and electrical design codes to which they are built must be specified to assure installation codes are followed. Asia, Europe, and Canada adhere to very similar electrical code requirements as does the United States and Central and South America. Australia and New Zealand are similar to European standards having some clear differences, thus one should request applicable standards to avoid inspection acceptance issues.

Direct current rectifiers have two standard DC power-related safety devices, thermo sensors on the thyristors and diodes, and thermo sensors in the transformer windings. Other safety devices include fan pressure switches, oil temperature sensors, oil level sensors, water flow sensors, and water temperature sensors.

Thyristor sensors provide an adequate safety margin to assure that the equipment is not operated too close to the upper temperature limits, usually 125°C. These devices are very reliable when design parameters assure normal operation is at least 25% less than the maximum temperature recommended by the manufacturer.

Diodes are able to operate at a temperature of 150°C. The same rule of thumb applies to diodes as well; don't design diodes to operate at temperatures more than 75% of the maximum operating temperature rating.

Air-cooled DC rectifiers are sometimes supplied with a fan pressure switch as a substitute for a vane type air flow switch or a pitot tube. Pressure switches can only be installed in equipment having an internal baffle at the diodes or thyristors to assure sufficient pressure. This pressure is a differential pressure measurement referenced to atmospheric pressure. Experience has demonstrated the necessity for a spare differential pressure switch available on-site. These switches are measuring pressure differences of $1/4$- to $3/4$-inch water column against atmospheric conditions, are in generally harsh environments, and are prone to develop sticky diaphragm failures over extended operating times. Cooling fan motors and oil or water circulation pump motors are always supplied with their own safety circuits in accordance with local code requirements.

5.8 Control panel

Control panel equipment can be classified in two different categories, relay logic based and programmable logic controller based. While relay logic control system faults are usually more easily located and corrected in the field, today these systems are being used much less. This change is the result of widespread use of distributed control systems where complete system integration requires sophisticated field signal instrumentation. Also, because of the increased use of powerful intelligent screen equipment allowing full system status monitoring, these changes are evolving much more rapidly. For this reason and because of the rapid development, acceptance, and use of computer based systems we will discuss programmable logic control systems having all the control features necessary for proper system control.

5.8.1 Instrumentation

Status, control, and alarm condition requirements for systems should include some or all of the following on each installation:

- Pressure and differential pressure control, sensing, indication, and pressure relief
- Liquid flow control, sensing, indication, and alarm
- Temperature sensing, indication, and alarm
- Tank level control, sensing, indication, and alarm
- DC rectifier operating status
- DC rectifier independent alarm conditions
- Dosing system operating control and status
- Chlorine residual analysis, control and indication
- Hydrogen detection and alarm conditions

5.9 Pressure and differential pressure equipment

On-site hypochlorite generation systems do not require sophisticated pressure indication (e.g., pressure transmitters and transducers) or controls. Pressure gauges used throughout the industry provide sufficient operating information for most applications. Gauges used in all except potable water applications should have diaphragm gauge seals to protect internal gauge components from liquid-borne corrosion associated with salt water and sodium hypochlorite. Gauge seals are assembled, vacuum evacuated, and filled with glycerin or silicone oil. They must then be bench-calibrated to assure accuracy. The use of oil filled gauges and stainless steel case materials provide excellent service life at minimum cost.

Situations arise in which the need for pressure control is essential to the safety of the system. Plant operations having very high water supply pressure will require at least a spring-operated diaphragm pressure regulator, a pressure control valve system or, at a minimum, an in-line orifice designed to create pressure reduction at required system flows. The diaphragm pressure regulator and pressure control valve will not allow the pressure to exceed a preset value. This pressure limitation will protect piping and equipment throughout a specific operating range. System protection is not part of plate orifice pressure drop systems except when there is design flow through the system. Plate orifice systems allow line pressure equalization at the highest available pressure when the flow is stopped. For this reason, the system water flow isolation valve should be placed on the high-pressure side of the orifice arrangement. Valves in this application are often tapered plug valves that allow good closure at high pressure.

Differential pressure (DP) devices are used on strainer systems in seawater applications for strainer cleaning. Differential pressure switches act as a control scheme override when the strainer receives an unusual filtration load before normal cleaning cycle operation is scheduled. Switches in this application are available with and without DP indication. One may use a DP transmitter, rather than a switch, connected to a PLC controller for indication and control. Those with DP indication must be installed where environmental conditions are not severe or one must design them into a protective panel. As with pressure gauges, DP instrument protection and calibration using oil-filled diaphragm seals is recommended to assure excellent instrument service in the field.

Pressure relief equipment is applied occasionally on brine electrolysis systems and nearly always to seawater electrolyzer equipment. With the exception of the ELCAT tubular cell design, a means of pressure relief is essential to protect cell equipment from excessive pressure. Particular attention must be given to the materials of construction to assure that corrosion will not cause loss of the relief device's safety function.

The two most common pressure relief devices employed in electrolysis systems are spring-loaded relief valves and rupture discs. PVC relief valves employed on seawater systems are not designed to provide full-flow, therefore it is necessary to provide two valves in parallel when full-flow is a requirement. Large metal valves, while not often used, can be sized to meet full-flow requirements. In each case, the material must be lined to protect against salt solution corrosion. Rupture discs are rarely used because they are one-time devices. If, for example, a high pressure water line upset occurs that causes disc failure, the system will stop and an extended shut down ensues to replace the failed disc. Continued operation is an advantage of the spring-loaded pressure relief valves.

5.10 Liquid flow equipment

Liquid flow control is essential to system performance, flow sensing to system safety, and flow indication to assure one of the essential conditions for maximum system performance and efficiency.

Though flow control devices vary widely, simple schemes are sufficient for effective operational control. Brine electrolysis systems are dominated by regulated pressure water supplies utilizing globe or tapered plug type control valves on the water and brine flow circuit. The importance of accurate pressure control must be emphasized to assure proper flow regulation. In fact, a pressure regulator for water flow control has been the most effective where system water pressure is unstable. This method of control works well when an ejector is used to draw brine solution from the brine dissolver for stable water pressure and accurate brine flow control as well.

Seawater systems are not as bound by close flow control requirements, therefore multi-orifice flexible flow controllers are used. These controllers employ many small rubber orifices installed in a plate and are accurate to ± 15% with the emphasis on plus. In situations where it is important to control flow more closely to a specific requirement, it is necessary to tune the orifice equipment. This tuning is accomplished by adjusting several of the small orifices to a larger or smaller size to achieve the desired flow rate at accepted operating pressures.

Flow sensing is normally accomplished by using paddle or vane type sensors. Where there are small flows of clean water (e.g., small brine electrolyzers), Hall Effect flow sensors are applied.

Small paddle sensors for water or brine flow using micro switches to indicate a condition change have the most utility for electrolyzer systems. These sensors are built with materials that will withstand both brine and hypochlorite environments, can be installed in very tight spaces, and provide sufficient accuracy for good safety protection. Applications include water, seawater, brine, and hypochlorite streams. Construction consists of Hastelloy, Viton, Ryton, and plastic body components. Sensors of this type are commonly applied to seawater feed circuits, electrolyzer cell outlet piping, hypochlorite dosing systems, and the like.

Vane sensors most often used are part of a variable area flow meter and applied almost exclusively to seawater feed piping. In this sensor, the vane is mounted to a cross shaft and flow passes through and beneath the vane. Mounted to the vane shaft are an indicator pointer and a cam unit. Inside a side mounted enclosure is a roller type side mounted micro switch. Using the cross shaft mounted cam to actuate the micro switch, a signal can be sent to the control panel to indicate operational status. These same vane sensors are available with analog signal outputs for use in control or as a PLC screen flow indicator.

Hall Effect sensors use a magnetic rotor spinning at a rate proportional to the flow. The rotor magnets cause a series of magnetic fields exciting the Hall Effect sensor. The pulse frequency is compared to a preset adjustable frequency activating the internal relay for a make or brake condition. This signal can be an RFO type 4.5-V to 24-V pulsed output signal or an RFA type 0–10 VDC analog signal to the PLC, then used to display the flow rate on a screen display. These small sensors are used predominantly on the water feed in small brine electrolysis systems. They are not as robust as is desirable for most field applications.

Flow indication generally consists of either a variable area flow meter or digital flow indication utilizing a signal from a Hall Effect sensor or an analog signal source. Variable area meters are the most commonly used being either tubular or an in-line paddle type arrangement. Because hypochlorite systems do not require extremely close control, accuracy of 5% is acceptable.

Alarm signals are achieved for most system designs by providing either a micro switch or a proximity switch contact closure. Where Hall Effect switches are used, the PLC is programmed to alarm when the signal is below a preset value. Micro switches are used in vane type variable area meters while proximity switches are most often applied to tube type variable area flow meters. All of these signals can be used with a PLC controlled system; however, Hall Effect devices must have an installed contact set for use in relay logic controls.

5.11 Cell level and temperature

All electrolyzer cells operating in a brine electrolysis environment will have cell level and temperature sensors. The level sensor provides assurance the cell is filled before applying DC power to the cell. The temperature sensor acts as a redundant sensor should the flow monitoring equipment fail to operate properly.

Level sensor failure or bypassing the electrical safety controls will cause premature DC power supply start up. If the electrodes are not completely covered there will be excessive liquid heating and undesirable bare electrode gas exposure. This situation will not normally be a problem; however, there is always the possibility for a spark occurrence between the electrodes. Because the electrolyzer always has some oxygen in the hydrogen mixture, the result of dissolved gas and anode inefficiency, this unflooded condition has the potential to ignite should any sparking occur in the cell.

The temperature sensor is further assurance that flow through the cell is sufficient to prevent cell and piping damage by sensing the cell exit liquid temperature. If the cell were made from titanium, then there would be no melt down; not so with plastics. Plastic cell bodies normally used will slowly fail at temperatures above 180°F (82°C).

5.12   Water and brine instrumentation

Water supply hardware feeding electrolyzer cells is either a small pump or flow meters with control valves using residential water pressure. Small pumps are used where the system is very small and stable water flow control presents a problem. Small diaphragm or bellows pumps provide the necessary control to assure that a small system, less than 50 lb/d (22.8 kg/d) will perform properly.

Small pumps must be chosen carefully. Local conditions having very cold water, less than 50°F (10°C), will cause premature failure of the bellows assembly in a bellows pump. In this case, a diaphragm pump is the best choice because diaphragms do not have severe flexing associated with bellows assemblies.

Flow meters for large systems normally are acrylic materials allowing corrosion resistance and cost effective installation ease in PVC piping. These meters are available with integral proximity sensors connected to the control system. This system assures that cell power will be stopped to prevent cell damage due to low water flow conditions. Low flow conditions will cause cell overheating and could cause a fire if electrode short circuits occur.

Flow monitors are employed on water feed systems where flow meters are not available with proximity sensors. Flow meters for flows less than 2 gal/min (8 l/min) require external sensors to monitor the water flow, again to prevent potential cell damage noted above. Brine flow control is normally by small positive displacement pumps to assure good flow control.

5.13   Inlet seawater strainers

Redundant strainer arrangements are normally required for large plants to assure reliable electrochlorinator (EC) operation. The use of both automatic and manual strainers is the most common. Engineering companies prefer automatic strainers to reduce cleaning maintenance and manual strainers in a parallel line to allow maintenance of the automatic unit.

Seawater systems require strainer assemblies for size control of seawater borne suspended solids. Particle sizes considered small enough to pass through a seawater electrolyzer must be less than 0.030 in. (0.8 mm). Some locations will experience an occasional plankton bloom. While this small aquatic animal passes through the recommended strainer opening, plankton will plug automatic strainers by their sheer numbers.

The separator media is either screen material in a supporting basket structure or wedge wire assembled to provide a straight tube type assembly. Wedge wire separators, being structurally stronger, allow less outside support. However, the supporting arrangement functions to hold the spiral wound wire at the proper position to maintain correct separator size requirements. Wedge wire assemblies also provide a very smooth surface for the scraper assembly to remove captured material built up on the surface.

Strainers are available having both manual and automatic cleaning capabilities. Materials of construction used in seawater are cast iron, brass, Monel®, copper-nickel, 316L stainless steel or related alloys 25–6Mo and Al-6XN and PVC. Fiberglass reinforced plastic (FRP) and high order metals such as titanium are used very rarely.

Basket strainers must be opened and cleaned manually through a gasket-sealed removable top. Bronze and stainless steel are preferred body construction materials. Seldom are more exotic materials used in body construction. Cast iron also serves well in basket units when proper cathodic protection is installed. Baskets are most often Monel or a stainless alloy.

Large, cast iron basket strainers have been used on seawater electrochlorination plants with good success. When basket strainers are paired with automated strainers the material of construction is usually stainless steel. Offshore platform EC applications are less susceptible to large quantities of suspended material. There, parallel PVC bodied strainers are used with good success, having one operational and one standby using Monel or stainless steel baskets.

Candle filters have been used where large seawater volume filtration is required. These are also manually cleaned arrangements. The body assemblies are made from cast iron and designed with upright filter units mounted to the assembly false bottom. Filtered seawater flows from the center of each candle into a collection area beneath the false bottom and is piped out of the casing bottom to the point of use. Separated material is dropped to the false bottom and removed from the side of the filter assembly. This type of straining device is not often used due to manual maintenance requirements.

Automated strainers are supplied in three basic design configurations, the spiral scraper, straight bar scraper, and plunger scraper types. Each design has been used successfully for seawater EC systems.

Spiral scraper and straight bar scraper designs are the most frequently used configurations. Flow rate sizing for all automatic strainers should allow for a 10 to 15% blow down rate. This assures that your system is not starved of seawater during normal operation.

Electrical control panels are supplied as part of these strainers providing a systematic cleaning cycle for operational maintenance. A differential pressure switch is normally connected across the strainer inlet and outlet to provide random cleaning, which is necessary only when unusual events cause strainer media blockage. This switch is electrically connected to the cleaning cycle timer controlling these random backwash cycles.

Gear reducers and motors are required for slow speed scraper operation. Scraper speeds of 2 to 4 r/min are used to remove the collected material from the strainer assembly. Removed material is then discharged from the bottom of the outer casing through an automatic valve to a drain.

Plunger designs are becoming more accepted in EC applications as they prove their reliability. This strainer design uses an air driven piston to push trapped material off of the wedge wire media. As with spiral and straight bar strainers, the piston can cycle nearly continuously in very severe deposit conditions. Removed material is scraped to the strainer body bottom and discharged via an automatic valve.

Control panels designed for piston scrapers are operated using air control logic. Differential pressure switches are not often used since the design is for the strainer to cycle continuously on a regularly timed basis. Frequency is adjusted in the control panel to meet requirements of the environment.

5.14 Brine system water softening

The water softener is used in brine electrolyzer systems to prevent calcium and magnesium hardness salts from depositing on the negatively charged cathodic electrode. As will be discussed later, systems can be operated without softened water; however, the potential for electrode damage becomes highly probable unless frequent stringent maintenance procedures are followed.

Water softening via ion exchange utilizes the replacement of calcium and magnesium ions in the water by an equivalent number of sodium ions. This system eliminates the undesirable characteristic of calcarious deposits because sodium salts do not form hardness type scales on the cathodic electrode.

The exchange ability of a resin is controlled by the total number of exchangable sodium ions. This exchange ability is referred to as resin capacity. When calcium and magnesium ions have replaced all the available sodium ions on the resin, the resin is then referred to as exhausted.

The total capacity of the resin, only present in the resin's full sodium form, requires a large excess of brine for each regeneration cycle. For this reason, a more economical volume of brine is used to obtain a reasonable operating capacity. In this respect, the regeneration efficiency, i.e., the total amount of hardness removed per pound of salt used, is important to minimize total salt usage for an on-site hypochlorite generation system.

The water softener contains a cation exchange resin (negatively charged), which is the same as used in home water softeners. Typical resin data are shown in Table 5.2.

Table 5.2 Ion Exchange Resin Properties

Structure	Crosslinked styrene/divinylbenzene
Form	Spherical Beads
Screen size, U.S. Std.	16–40
Swelling	5%
Water retention	45–47%
pH Range (stability)	45–48%
Approximate weight	52 lb/ft^3
Nominal capacity	30,000–35,000 grains
Chlorine maximum	1 ppm

Water hardness may be expressed as grains per gallon. One grain per gallon is equal to 17.1 parts per million (ppm) per grain. For example, water with 10 grains of hardness will be equal to 171 ppm.

Regeneration of the ion exchange resin is achieved by drawing some of the brine solution from the electrolyzer salt dissolver. Each regeneration valve has a brine suction ejector which is properly sized by the on-site generator manufacturer or regeneration valve manufacturer to provide the correct amount of salt to assure the resin achieves its maximum operating capacity.

Two types of water softeners generally available in the marketplace that are used with brine electrolyzer systems are single and double tank water softeners.

Tanks are fabricated from steel or FRP. Large tanks of steel construction remain in some situations, often with a liner material to provide long-term corrosion protection. The sizes required for on-site hypochlorite generator systems are now fabricated from FRP.

Regeneration valves are either electrically or hydraulically operated. Electrically operated valves will use either a water volume usage sensor or a 24-hour tab type timer to initiate the regeneration cycle. Hydraulically operated timers use only a water volume usage sensor for regeneration cycle initiation.

Materials of construction for regeneration valves consist of brass and molded plastics. Many manufacturers are moving away from brass to eliminate the leaching prospect of the alloying materials in the treated water. Plastic material such as Noryl® is often used due to its molding ease and accuracy.

Both single and double tank style units are available with electrically operated 3-cycle or 5-cycle regeneration valves. The 3-cycle valve has brine, slow rinse and service functions while the 5-cycle valve has backwash, brine/slow rinse, fast rinse and service functions.

Single tank equipment will expose the electrolysis cell to some hardness. This occurs because the valve bypasses raw influent water during the regeneration cycle. As a practical matter however, either valve is satisfactory for use on the brine electrolysis systems since the water hardness exposure time for the electrolytic cell when using single tank equipment is low, approximately 3% of the total operating time.

On-site generator manufacturers provide the water softener with their equipment to assure the customer will achieve long reliable cell electrode service from the cell electrodes. Softeners also reduce the need for chemical cleaning since a significant portion of the calcium material is resident in domestic potable water rather than the salt used in the electrolysis process.

An example of water softener importance is as follows. The water has 100 mg/l hardness and the salt has 1500 mg/l for the cell feed material. The salt is dissolved to a 30% solution to form a solution hardness 550 mg/l (1500 mg/l · 0.3 + 100 mg/l). This solution is then further diluted by a 10:1 ratio. Therefore, the salt contributes 55 mg/l (1/10th of 550 mg/l) and the water contributes 100 mg/l. As you can see above, the electrolyzer cell is exposed to 3 times the hardness when no water softener is installed in an electrolyzer system.

5.15 *Temperature sensing equipment*

Temperature sensing falls into two areas of a hypochlorite generation system, cell discharge liquid temperature and DC rectifiers. Direct current rectifier temperature sensing and control will be discussed in the DC rectifier section of this chapter.

Cell liquid temperature is used as an independent back up safety circuit for the flow sensing equipment in brine systems. Temperature indication is seldom used in brine electrolyzer cells primarily because the flow rate in brine systems is quite sufficient to prevent excessive temperature rise in the cell. The use of only a cell thermo switch is required for adequate protection. Metallic thermo sensors must be protected with either corrosion resistant metallic or plastic coated thermo wells. This protection may create a sensitivity problem when the well delays the temperature ingress to the thermo sensor device.

Seawater systems have been installed with temperature switches and indicators. This equipment was protected as noted above with thermo well assemblies in the hypochlorite stream. Seawater systems operate at very high flows where the solution temperature rise across the system is normally less than 3ºF (2ºC), therefore the likelihood of a temperature related shutdown is extremely remote. For this reason few systems have installed temperature switches. The authors have experienced one seawater system shut down from high temperature. In that case a workman's leather glove was caught in the vane flow monitor causing low flow.

Another cause for high temperature in the cell that will normally not be recorded, is when the cell has not been properly acid cleaned (an essential maintenance procedure). In this situation, the cell will have a small channel in the deposits through which the flow passes at very high velocity. The electrodes are at risk of damage when this condition occurs. Deposits encapsulate the electrodes and damage both the anodes and the cathodes, yet there are no problem indications. This problem occurs frequently to those operators who do not maintain their cell properly despite the built-in safety features.

Systems are designed to shut down and alarm whenever there is a temperature related alarm. Sensor set points of 140°F (60°C) for brine electrolyzer cells are normal. Temperatures exceeding 140°F (60°C) put the PVC piping usually used in these systems at risk of sagging on poorly installed pipe runs and joint failure at elevated pressures.

Seawater systems are normally set to shut down and alarm at 105°F (40°C). This temperature is elevated sufficiently to provide for the temperature rise through the cells in the hottest climates. Seawater temperatures above 95°F (35°C) are very rare, thus the choice of a set point of 105°F.

5.16 Cell level and temperature

All electrolyzer cells operating in a brine electrolysis environment should have cell level and temperature sensors. The level sensor provides assurance the cell is filled before applying DC power to the cell. The temperature sensor acts as a redundant sensor, should the flow monitoring equipment fail to operate properly.

Level sensor failure or bypassing the electrical safety controls will cause premature DC power supply start up. If the electrodes are not completely covered, there will be excessive liquid heating and undesirable bare electrode gas exposure. This condition will not normally be a problem; however, there is always the possibility for a spark occurrence between the electrodes. Because the electrolyzer always has some oxygen in the hydrogen mixture as a result of dissolved gas and anode inefficiency, this unflooded condition has the potential to ignite if sparking occurs in the cell.

The temperature sensor is employed as further assurance that flow through the cell is sufficient to prevent cell and piping damage by sensing the cell exit liquid temperature. If the cell were made from titanium, then no melting would occur as with plastics. Plastic cell bodies normally used will slowly fail at temperatures above 180°F (82°C).

5.17 Tank level equipment

Tank level control schemes offer a variety of reliable level control devices. The circuit design for both brine systems and for seawater systems usually is designed with four set points, two for alarms and specific system operational conditions and two for normal operation as noted below.

Low–Low tank level — The hypochlorite generation system is able to operate while the dosing pumps are locked out to prevent pump damage from a dry running condition. This condition will cause an alarm, must be manually reset for the system to begin operation, and will clear automatically upon reaching the Low level.

Low level — The system is fully functional. This level is the system starting level where hypochlorite generation is initiated to refill the tank and the dosing pumps are fully operational. This level is at least 6 in. (152 mm) above the dosing pump suction outlet for flat bottom tanks and much greater for coned or radius bottom tanks.

High level — At this level, the hypochlorite generation system operation is stopped and the dosing pumps are operating or operational depending upon the dosing scheme.

High – High level — This level should be below the overflow line level, override the high level shut down control, and provide a condition alarm. The pumps should be operating or have an operational fault.

Systems designed in the early days of on-site hypochlorite generation used air operated differential pressure systems where either pressure switches or cam actuated micro switches provided set point control. The pressure switch circuit would incorporate a face-mounted level indicating gauge in the control panel door with the air controls and internally panel-mounted pressure switches connected to the control logic. Cam actuated arrangements used a bellows operator to indicate the level and switches within the body also connected to the control logic. The control logic would be set to cause an alarm should the tank level exceed a preset value. Float type level switches were also installed in some early systems. Some materials of construction issues caused reliability problems that were easily overcome by using air operated systems.

Today's level control units are more likely to be ultrasonic sensors, proximity switch arrangements, or differential pressure (D/P) switch designs. Ultrasonic level systems are very reliable if installed in an environment with minimum vapor condensation to cause corrosion and interfere with the unit accuracy; therefore, these sensors must have good air circulation. System PLC connections allow operational control and alarms for the system while the use of user interface screens can illustrate tank level.

Proximity switch systems offer very reliable operation. Mounting is on the side of the tank from flanges, allowing level indication locally and variation of the sensors along the level indicator member. Proximity switches provide contract sets for system operation only without the benefit of control panel level indication.

D/P switches are becoming the most frequently applied level control equipment. Differential pressure equipment offers many benefits for the control designer: accurate level measurement, 4–20 mA analog output, NEMA 4X weather rating, severe classified area compatibility, wide pressure range capability, and ease of tank connection. The reliability of D/P cells and the removal of air as a system operational requirement are becoming a requirement by engineering firms for field installations. Materials of construction are to have Hastelloy C diaphragms and bodies for corrosion prevention and a level range up to 250 in. (6350 mm) of water column head.

Some engineering firms may require a separate D/P device for alarm functions to reduce the possibility of a total system failure. Redundant equipment is attached to the same tank outlet connection with independent liquid isolation valves also with independent instrument cabling.

Differential pressure units are very cost effective when connected to integrated control systems in use today. A PLC used with computer integrated screen systems allows the designer to put the system flow sheet on the screen to indicate the tank operating level.

5.18 Salt storage-dissolver tanks

Salt dissolver systems are designed to meet a myriad of sizes, brine demands, and filling methods. Dissolver units for brine electrolysis systems are available in three designs to meet sizing needs. These units vary from tank type dissolvers to high capacity pit dissolvers. Smaller system dissolvers are sized for 3 to 7 days between refills while large dissolver systems are designed for a minimum of 15 days storage to minimize delivery requirements.

Dissolver tanks are normally built of polyethylene (PE), fiberglass reinforced plastic (FRP), or concrete. These materials of construction are defined primarily by tank size and the loading method required to accommodate the required electrolyzer system salt volume. Tanks of 2000 lb or smaller are nearly always installed indoors or under cover. Large FRP tanks and pit dissolvers will always be outside the electrolyzer building for easy loading access.

Small tanks constructed from PE or FRP are designed for up to one ton of salt. Tanks of this size require hand loading from 50-lb salt bags available from local suppliers on pallets. These dissolvers are usually used for systems smaller than 200 lb of chlorine per day because manual loading is required.

Large tanks are also constructed from FRP and sized from 12 tons to 100 tons using blower trucks to offload the salt into the tank. Bags mounted outside the tank are designed to capture the fine salt to prevent salt dust from causing local hardware corrosion. Units having this capacity are used with systems of several hundred to several thousand pounds chlorine capacity per day.

Pit dissolvers are used on projects where there is a very large volume of chlorine required. Loading methods vary widely from dump truck to rail car arrangements. These systems are common in plants where it is more convenient or more cost effective to receive bulk shipments. Pit dissolvers require almost no maintenance particularly if the salt quality is low in suspended solids, preventing bottom sludge buildup.

5.18.1 Dissolver level controls

Level control systems use float assemblies or electric conductance probes to control salt dissolver water addition. Float type systems usually use multiple ceramic float units assembled with a stainless steel wire suspended from the top of the tank and connected to a top works assembly housing the electrical switch arrangement. The level float arrangement is enclosed in an FRP pipe also suspended from the tank top. The floats are protected from the salt and modulate with the tank brine level. Mercury type switches are connected to an actuation assembly in the top works. These mercury switches connect to and actuate the inlet water control valve.

Conductance probes are inserted in a standpipe assembly on the outside of the tank via a flange near the bottom of the tank. There are three probes on the head assembly, a reference probe, a high level and a low level probe. Power conductance through these probes and the salt solution actuates the control relay operating the water control valve on the inlet water line. The brine dissolver is normally maintained at a liquid level, usually 2 to 3 ft of solution, immersed in the stored salt.

5.18.2 Internal distribution and brine removal piping

The purpose of the distribution piping is to maximize the salt dissolving rate. It is normal to have dissolving rates of 25 to 50 gal/min in tank type dissolvers. Pit dissolvers have very high surface areas, therefore their dissolve rates are much higher. Dissolvers are used where there is a need for high volumes of brine solution. These rates are not necessary in a brine electrolysis system because the feed rates are comparatively low, less than 10 gal/min.

While the tanks are FRP material, internal water distribution piping is normally PVC. Due to the piping arrangement, dissolver internal piping is normally field erected to assure damage-free equipment for start up operation of the dissolver.

Brine removal is accomplished through two major devices, using inverted hat type collectors or perforated pipe collectors across the bottom of the dissolver tank. Dissolver systems most often use mined granular or seawater evaporated solar salt. When granular material is used there is no need to prevent granular material from being drawn into the collector system since the salt acts as its own filter bed. In the case of refined purified salt it is very important to provide a barrier of gravel over the collector system. This gravel bed is placed in the dissolver, after all piping and distributors have been installed, to a depth of approximately 12 in. (305 mm). The gravel is graded in a layer of 3/8 – 1/2 in. (9.5 mm–12 mm) size over a layer of 1/8 – 1/4 in. (3 mm–6 mm) size material.

5.18.3 Salt addition

Salt addition methods are determined by tank design. Small tanks, up to 2000-lb capacity, are normally filled manually from 50-lb (23-kg) bags. Salt delivery in these cases is usually by palletized bags to minimize bag handling.

Large tanks are filled by another method because the quantity of salt required demands more efficient delivery methods. Most tanks used on large electrolyzer systems utilize blower trucks delivering 20 tons of salt each. The delivery line is usually 4 in. in diameter designed of aluminum having a smooth long radius into the top of the tank. The interconnection between the tank and truck is made using a cam-lock quick disconnect. The air is discharged from the tank through a line with a bagged end to collect the light salt dust blown out of the tank during the transfer procedure. Salt transfer is an easy operation; however, it is important to assure that both inlet and outlet lines always remain clear of obstructions. General maintenance procedures applying to salt dissolvers can be found in the appendix to assist you with frequently asked questions.

5.19 Product storage

Storage tanks allow the system to operate at full capacity to achieve optimum performance, provide hydrogen gas removal and when applicable, provide backup product supply during periods of maintenance down time. Materials of construction most frequently consist of high density polyethylene (HDPE), fiberglass reinforced plastic (FRP), or a concrete pit having an impermeable liner. High density polyethylene is preferred for smaller systems and FRP for large systems. Concrete pit storage is used more for very large installations where plant design allows convenient pit integration. The plant designers must consider freeze protection and take advantage of appropriate types of material during plant construction. For all installations where a freezing environment is possible, freeze protection must be provided to prevent tank damage in all cases except pit designs.

5.19.1 Tanks

High density polyethylene-small polyethylene (PE) and high density polyethylene (HDPE) tanks up to 500 gal (2000 l), with a removable cover, are configured with bulkhead fittings or hot air welded gusseted flanged nozzles for product inlet and outlet connections. The cover must vent to a location outside the building for hydrogen gas venting. Tanks of this size are seldom installed with forced air blowers for hydrogen dilution. Blower use is controlled by local fire code regulations, therefore regulations must be consulted to assure compliance.

Larger molded dome top tanks will have the same connection arrangements as noted for small tanks. Because these tanks have no removable top, a top-entry manway is part of the tank design to allow entry access for maintenance. Each tank type is offered in translucent and black finish. When using black tanks, installation consideration should be given to the solar heating potential of the product resulting in product degradation.

Fiberglass reinforced plastic — Fiberglass tanks are typically used for storage requirements of greater than 5000 gallons (19,000 liters). With few exceptions, FRP tanks are custom built to suit each site's specific requirements. The tanks are designed with flanged nozzles to accommodate product handling, hydrogen dilution fan ducting, a side and/or top bolted access manway, and anchor lugs around the bottom of the tank for mounting able to withstand wind loads of 125 mi/h (210 km/h).

FRP tanks used to store sodium hypochlorite product should be manufactured to ASTM D-3299, latest revision to assure tank quality. The tank vendor must supply a 20 mil vinyl ester resin rich Nexus vail liner that is BPO-DMA (benzoyl peroxide-dimethylaniline) cured as specified for sodium hypochlorite service. Exterior tank material may be the less resistant isophalic resin. A white gel coat should be applied to assure good UV resistance, and, in climates where the tank is exposed to freezing conditions, insulation and heat tracing must be added during the manufacturing process.

Pit storage — Pit storage is facilitated at the time the plant is being designed. Since concrete is the material of choice, sodium hypochlorite and salt corrosion properties require this storage facility be installed with a flexible lining material. The liner provides corrosion control for the structural reinforcing materials supporting the concrete. Liner material applied to sodium hypochlorite storage is a plasticized PVC. This is available in 20 to 80 mil thickness having PVC's corrosion resistance with the added benefit of flexibility for easy installation.

Tanks or level controls require no maintenance unless there is an unusual catastrophic failure. HDPE tanks are usually replaced while FRP tanks may be field repaired. When repairing an FRP tank one must assure the repair materials and procedures are suitable for the service application or one's courting a future failure.

5.20 Pump equipment

5.20.1 Seawater booster pumps and seawater hypochlorite dosing pumps

Pumps to meet seawater supply and dosing requirements are always centrifugal designs. For U.S. EC system manufacturers, the choice of an ANSI B 73.1 pump design is the most suitable to facilitate interchangeability between pump manufacturers. Cast iron, rubber lined steel, stainless steel, and FRP are the most common materials of construction for seawater supply requirements. Dosing pumps are limited to rubber lined steel, FRP, titanium, and Hastelloy C due to the corrosive nature of sodium hypochlorite. Plastic materials such as PVDF and PVC have been used for EC system applications with good success. Most engineering firms are reluctant to accept plastics, however, because most large sites are overseas, requiring heavy spares stocking to assure reliable EC unit operation.

Today materials of choice for seawater handling requirements are normally FRP and stainless steel. The initial pump cost may be slightly less for cast iron and rubber lined steel, while maintenance requirements are potentially greater for pumps with plastic construction materials.

Recently, magnetically coupled pumps of sufficient volume and head capacity have been introduced which eliminate seal leakage problems associated with more traditional designs. Seal assemblies are chosen to avoid the need for seal flush water systems when sites are remote from acceptable flush water supplies. Internal flush arrangements are acceptable when the pumped material is seawater or hypochlorite. The proper materials choice is a ceramic stationary and a rotating carbon unit housed in Hastelloy or titanium case and springs with Teflon, Viton, or EPDM seals. Care must be applied to the carbon seal choice to confirm compatibility with hypochlorite solution. Field experience has demonstrated that seals with these materials of construction will withstand the environmental rigors of seawater and hypochlorite exposure.

Motors should be of TEFC (totally enclosed fan cooled) design. This system of motor enclosure assures water ingress problems are not visited upon you soon after the system is operational. Also, with rare exception, TEFC motors are accepted for NEMA classified areas with possible temporary explosive gas exposure such as NEMA Class 1 Division 2 Groups C&D areas. Continuous gas exposure areas such as Class 1 Division 1 Group A&B will always require explosion-proof motors.

Pump base assemblies can be an issue where the base is nonmetallic. Nonmetallic bases must be properly grouted in the field to assure pump stability in operation. Low level pump vibration over a long time period will cause eventual structural failure of an ungrouted nonmetallic base. This problem is particularly true when you are using very large pumps and motors.

Couplings and coupling guards are supplied with these pumps as a standard practice. Coupling alignment is completed at the pump supplier facility, therefore field alignment should only be necessary when the motor or pump is removed from the base. Alignment checking is then only necessary to confirm the pump and motor are within acceptable alignment tolerance limits.

Coatings on steel pump bases, motor housings, and coupling guards should be suitable for the expected environment. Cast iron motor housings will withstand most severe conditions regardless of the type of factory coating. Steel motor housings, base assemblies, and coupling guards are most often neglected. Painting with a 3- to 4- part Carbolene of Ameron coating system specified for offshore duty is recommended.

Pumps of choice for seawater and hypochlorite in the United States are FRP material of construction. FRP pumps meet specification ANSI B73.1, are applicable on either seawater or hypochlorite, and are vinyl ester resin having a BPO-DMA (benzoyl peroxide-dimethylaniline) cure to assure high resistance to either pumped materials.

Pumps manufactured from the various plastics other than FRP are rarely used in these high volume sizes. While PVDF dosing pumps offer excellent reliability, they have been used only on rare occasions since this material is more expensive. PVC or PP pumps have also been employed although the larger sizes are not as reliable in large volume extended operation.

Metallic materials such as titanium or Hastelloy C are also not often applied in seawater systems. On rare occasions where there is a need for very high discharge pressure these pumps will be used. Most of these applications are the requirement of an engineering firm or an end user where metallic pumps offer a comfort level over alternatives.

Magnetically coupled pumps have been used for very small applications very successfully. Data on large volume magnetic drive pumps indicate similar effectiveness for these as well. Pricing is a consideration for these pumps when compared to conventional pump drive designs, thus use is presently limited to special situations or requirements.

In summary, all of the pumps defined will function well when chosen for the application carefully. Care in choosing the materials of construction and the seal design will assure excellent performance life. The authors believe developing vendor lists for more cost effective magnetic drive pumps will gradually shift this configuration as the design of choice in seawater electrolysis applications.

5.20.2 Brine system product dosing

Brine electrolysis systems will generally have diaphragm or hose type positive displacement pumps serving chlorination dosing requirements. These pumps do not have flow sensors to confirm dosing operations are functioning properly. Dosing control and monitoring are carried out by the system's residual analysis equipment outputting an analog feedback signal. Systems used for water well disinfection will not normally have any residual analysis because the well operates at a constant flow, therefore the dosing pump has a set and forget condition.

Unusually large systems will have centrifugal pump dosing systems. When centrifugal pumps are used in a system, flow monitors may be applied. Such pump systems are designed to have a flow monitor where the level of output is set greater than 35% of pump rated flow to prevent pump damage. A centrifugal pump's output control is normally via a discharge control valve in turn controlled by the analog signal from a residual analyzer. Care must be taken in the design phase to assure that the system is not so oversized to cause nuisance pump shut off from low output flow. Standard motor controls meeting local electrical codes are used to protect the drive equipment in all pump circuits regardless of pump design.

Diaphragm type positive displacement pumps — Diaphragm pumps are most often used in dosing systems having less than 200 lb (90 kg) chlorine capacity per day. The corrosive properties of sodium hypochlorite require pump heads manufactured from PVC, acrylic, or PVDF utilizing Viton or Teflon diaphragms. Although the product utilization tables suggest that EPDM is an acceptable material for this application, experience has proven that EPDM is not a reliable diaphragm material. Ceramic check valves must also be utilized to be confident of optimum pump performance.

Peristaltic hose type positive displacement pumps — Peristaltic pumps operate on the principle of rollers collapsing a hose to create a vacuum drawing in fluid and discharging at elevated pressures. These pumps utilize variable speed motor controls with either VFD or SCR drive controllers. No check valves are required since the roller position acts to prevent fluid backflow. Pressures of 230 psi (16 kg/cm^2) can be achieved when using these pumps. The hose assembly on larger pumps has raised face flanges for ease of piping connection and a front cover assembly facilitating hose exchange. One should always be aware of flange end and hose end support sleeve construction materials to avoid corrosion. <u>When sodium hypochlorite is the application, titanium is the material of choice.</u>

Centrifugal pump designs — Centrifugal pumps are only considered where large product volumes are required. Pump materials should be PVDF or vinyl ester resin fiberglass with a catalyst cure specified for use in sodium hypochlorite. Shaft sleeves and seals must be fully compatible with sodium hypochlorite and can be designed for either external water flush or internal product flush regimens.

Recent improvements in magnetically driven pumps assure this design is reliable by eliminating the potential for seal leaks. Be reminded however, the body and impeller construction materials must be compatible with sodium hypochlorite.

Unlike positive displacement pump designs, external flow control schemes are necessary to assure proper dosing control or level control. These schemes are accomplished with flow control valves operating from an analyzer, flow indicator, or level control arrangement. Instrumentation designs utilize an analog PID control loop. This control loop is connected to an I/P (instrument to pneumatic) transducer air output providing a 3–15 psi output to air valve motor operators or directly to an electric valve controller motor accepting an analog control signal.

Pump controls — Manual control modes are used primarily for domestic water well applications with a fixed flow rate and dosing requirement. Small positive displacement pumps are used with a micrometer type graduated adjustment in this application. Pump controls are configured to operate only when the well pump is in operation, whereas an on-site sodium hypochlorite generator will operate independently when the storage tank level controller indicates tank refilling is necessary.

Plant water flow instrumentation provides an analog signal to PID loop controllers pacing dosing pump speed or dosing valve position for water system dosing flow control.

A continuous sampling residual analyzer will output an analog signal to a PID loop controller pacing dosing pump speed or dosing valve position for water system dosing flow control.

Integrated pacing dosing controls combine both plant flow and residual analysis. Plant water flow and residual analyzer analog signals are integrated using a PLC algorithm to define the trend, usually on a feed forward basis, then output an analog signal to the dosing pump or valve arrangement for dosing feed rate control.

5.21 Pipe, valves, and fittings for electrolysis systems

5.21.1 PVC (polyvinyl chloride) and CPVC (chlorinated polyvinyl chloride)

Polyvinyl chloride (PVC) is the most cost effective material in all areas of brine and seawater electrolysis. Chlorinated polyvinyl chloride (CPVC) is less desirable, as previously noted, for cost reasons and the more brittle nature of this compound.

Valves for these materials may be ball, butterfly, diaphragm, gate, globe, needle, etc. Seal materials should be Teflon, Viton, or EPDM to assure good service life. Where there may be severe deposits on the valves, ball valves and butterfly valves should be avoided. An example of such a site would be high salt brine wells where the deposits consist of calcium carbonate rather than magnesium hydroxide experienced in seawater. Deposits will destroy the seals, rendering the valves ineffective. In these rare cases the valve of choice is the weir type diaphragm valve with a Teflon faced diaphragm assembly.

5.21.2 Polypropylene

Although this material is applicable for some portions of a brine electrolysis system, its use is only recommended for water and brine handling on brine systems. This material used as tubing and fittings is acceptable as it is the most cost effective.

5.21.3 Acrylonitrile butadiene styrene (ABS)

This material is, as is polypropylene, only recommended for use for water, brine, and seawater applications. Although the authors have seen this material used in European and Asian supplied cell interconnection equipment, a U.S. manufacturer will not recommend this material for reliable service in sodium hypochlorite.

5.21.4 Polytetrafluoroethylene (Teflon®)

Kynar and Teflon are applicable throughout the system. Cost is the main issue for this material, therefore use is generally limited to pump heads, pump diaphragms of Teflon, and pump check assemblies utilizing ceramic ball material. Both materials are tolerant throughout the application spectrum in brine and seawater electrolysis.

5.21.5 FRP (fiberglass reinforced plastic)

FRP is also an expensive material of construction. The authors would not utilize FRP for on-site hypochlorite systems except where it is specified or in unusual environmental or pressure conditions. High quality butterfly valves are available and are used in situations where pressures can exceed PVC pressure capacities when the temperature is elevated.

5.22 Hydrogen handling practices

Hydrogen is produced on all on-site sodium hypochlorite generators. As you have seen in the equations in Chapter 4, hydrogen is a by-product of the electrolysis process with virtually 100% electrolytic efficiency at the cathode.

As hydrogen is produced in the cell it combines with dissolved air and oxygen generated from the chlorine inefficiency of chlorine production at the anode. The undiluted gas exiting the cell is approximately 3% air and oxygen, therefore is below the lower explosive limit (LEL) of 4.0% in an air environment. The explosive range of hydrogen with air is between 4.0 and 74.20%. In an oxygen environment the explosive range is 4.65 to 93.9%.

Air dilution requirements are dependent on the system output and tank type. Although one cell design separates the hydrogen within the cell, allowing its hydrogen to be directly vented to atmosphere without the need for air dilution, all others include the hydrogen in the product stream to the storage tank.

All brine electrolysis systems use storage tanks having closed tops, either as loose or fixed covers, concentrating the hydrogen in the gas void area above the liquid product. Without exception, storage tanks with closed tops require dilution air to assure that this gas void fraction is below the hydrogen LEL. To achieve low LEL levels the system should have air blowers installed that will dilute the incoming hydrogen by 50:1 to 100:1 with air.

Blower duct arrangements must be designed to assure that the air has unrestricted flow into the tank. Where multiple fans are used the ducting must have an air check assembly to direct the air into the tank rather than out the adjacent blower's suction opening.

Open top tanks are used in seawater electrolysis systems where plant locations are remote from possible ignition sources. In addition, hydrogen in this application separates upon tank entry above the solution level dispersing into the atmosphere very rapidly. High dispersal rates are the result of the difference in the specific gravity of hydrogen versus air, 0.0695:1, or .00523 pounds per cubic foot, the weight density of a cubic foot of hydrogen.

Although hydrogen solubility in hypochlorite product is low, it is important to remember that sodium hypochlorite generated on-site will be hydrogen saturated. At a temperature of 25°C and atmospheric pressure the theoretical hydrogen absorption at saturation is 0.0001535 g hydrogen per 100 g of water. Because the solution is warm, 40 to 50°C, hydrogen solubility is elevated, therefore as the solution cools additional entrained and dissolved hydrogen will gradually release from the solution into the tank gas void space. Normally this is not dangerous since the tank should have an atmospheric vent to allow released hydrogen to escape where no dilution blowers are used.

Seawater systems also utilize a hydrocyclone system taking advantage of the high separation efficiency caused by the extremely low hydrogen specific gravity.

Seawater systems with high production capacity also have high seawater feed volume requiring a large inlet and outlet pipe connection size. A large pipe connection size does not affect separation efficiency. Product streams from a group of cells will have as much as 50% entrained hydrogen as small bubbles, yet the hydrocyclone will separate 95% of this entrained gas. By pooling the liquid in the hydrocyclone cone with a vortex breaker in the discharge line, the vortex breaker prevents separated gas re-entrainment into the discharge stream.

Separated gas is vented through a small titanium orifice from a top discharge connection. The separated hydrogen is piped through a liquid seal pot assembly for direct atmospheric venting or into a large diameter vertical pipe. The vertical pipe will have dilution air blowers attached to dilute the hydrogen below the LEL within the pipe before venting.

The hydrocyclone provides a convenient method of using the seawater cell feedstock pressure as a form of dosing arrangement. Backpressure from the dosing piping to the dosing point will balance the pooling level in the hydrocyclone to avoid flooding, i.e., there is always a hydrogen head above the liquid regardless of the system production level. As noted above, several methods may be employed to dispose of the waste hydrogen generated during on-site sodium hypochlorite production.

Use of a storage/degas tank with attached air blowers is the most common practice. For this scenario the tank serves as storage and as a pump head tank for continuous and shock dosing regimens commonly applied to large circulating cooling systems. The control schemes operate the EC system to assure proper tank level–pump discharge coordination. Air blowers are sized to dilute the hydrogen to below the lower explosive limit, LEL, of 4 % in air. Here again, generally accepted practice is for 50 to 1 or 100 to 1 dilution.

When a small product degassing tank is provided, generally accepted practice is to provide at least 3 minutes of tank residence to assure sufficient hydrogen removal. This arrangement provides dosing pump suction head for applications where the dosing point pressure is too great to utilize a hydrocyclone. The controls provide proper tank level/pump discharge operation to avoid pump damage. The air blower sizing remains the same as with large tanks, 50 to 1 or 100 to 1 dilution.

A third hydrogen removal practice is to employ a hydrocyclone in the product discharge line. For this case no large tanks or dosing pumps are utilized. Seawater supply pressure is utilized to provide the necessary pressure to the dosing header. Because hydrogen's specific gravity is 0.06 that of air, separation is easily accomplished by the centrifugal force created in the cyclone. A pool of product is maintained above the discharge line by a vortex breaker baffle in the cyclone product discharge.

The separated hydrogen is passed from the top of the cyclone through a small orifice. The orifice opening is very small regardless of the hydrogen release volume requirements, having a size usually less than 0.188 in. (4.75 mm). There is always a small amount of liquid bleed through the orifice that must be released to a nearby drain. Liquid is released from the hydrogen discharge line by a small U trap from the branch of a pipe tee installed a small distance above the hydrogen discharge flange.

Hydrogen is directed to a water seal pot or a vertical pipe where air blowers dilute the hydrogen to the required dilution level.

When a seal pot is used there is only a need to vent the hydrogen to atmosphere well above any adjacent roof or above 15 ft (7 m) in an open area. A continuous flow, either seawater or drinking water, is maintained through the seal pot while the hydrogen is bubbled through 2 to 6 in. of the flowing water. This design will prevent a hydrogen ignition flashback.

Where hydrogen dilution is required, cell produced gas discharged from the cyclone enters a pad mounted vertical pipe that accepts the hydrogen and required blower air venting the combination to atmosphere. The blower sizing is consistent with previously noted practices.

5.23 Brine system applications

Sodium hypochlorite systems have been used in drinking water and wastewater applications since their early development. Electrolytic cell development in the early 1970s started the search for plants where this radically new concept would be accepted. The result was to sell a few drinking water and wastewater plants in the United States in the late 1970s with entrepreneurial managers operational systems.

These systems were successful in proving on-site systems as a viable alternative to chlorine gas for water treatment plants. The most severe obstacle was the cost of chlorine gas throughout the United States. Plant users could not operate this equipment cost effectively when compared to the cost of chlorine gas. To the authors' knowledge only one serious attempt to improve unseparated electrolyzer cell performance to mitigate the deficit with chlorine gas cost has been published in the previous 25 years. It was not until the advent of the EPA regulatory requirements in 1998, noted earlier in this book, that the cost of chlorine became a smaller issue.

Presently, water utilities are converting to on-site generation for small to moderate plants particularly where plant upgrades are taking place. This changeover continues to evolve slowly as the technology becomes more accepted and as the advantages of small systems installed at remote well sites or centralized in housing developments become evident. It is also useful to note that approximately 80% of all drinking water systems in the United States require less than 500 lb of chlorine per day.

5.23.1 Cooling system applications

Cooling water systems, both single pass and recirculating loops, have used large brine electrolyzer systems where having chlorine on site was not desirable, nuclear power stations in particular. These systems were installed in the late 1970s and early 1980s. The reduced cost of commercial bleach since this equipment was installed caused these plants to be decommissioned. The storage tanks installed with the system are utilized for commercial bleach storage and the pumps changed to a smaller size to allow for the higher strength product. Cooling tower applications have not been pursued heavily because large water treatment chemical companies such as Nalco contract for the tower care service, which includes the necessary chemicals. Electrolyzer cell developments just now being announced could well cause a return to on-site hypochlorite systems as a cost effective alternative to commercial bleach.

5.23.2 Food processing

Food processing of all types use some form of chlorine disinfection chemistry. Food processors, meat processors in particular, have a local USDA inspector with significant control of the plant disinfection process since they are responsible for plant bacterial conditions. On-site systems have been tested in chicken processing plants with excellent success, controlling the level of chlorine in the wash and chiller areas. In these cases, the chlorine level is tested multiple times per day to assure it is within defined parameters, usually between 20 and 35 mg/l. Besides meat, poultry and fish processors, fruit and vegetable washing operations are application areas as well. Acceptance in this area will gradually occur as the advantages of on-site equipment become clear.

5.23.3 Beverage operations

Beverage operations utilize various types of disinfection for bottle washing operations, beverage water, and equipment and plant cleanup. In the early 1980s Coca Cola – Japan used on-site sodium hypochlorite generators to disinfect in their bottle washing operations and bottle water. Very precise testing of the water was performed to assure that the bottle water met the strict standards required by Coca Cola. This equipment has also been applied in South America in smaller bottling operations. While this is a limited market the application is further proof of the diverse applications for on-site systems.

5.23.4 Water disinfection

Irrigation water disinfection, while not essential, allows much more reliable and longer drip system operation. The biological activity in most irrigation system wells will plug the pores of irrigation lines, then requiring these lines to be changed annually. Sodium hypochlorite is a common application for disinfection; however, sodium in the water will eventually change the soil sodium levels to a point that is very unfriendly to plants.

An alternative to sodium hypochlorite is to use potassium hypochlorite. Potassium hypochlorite offers the opportunity to disinfect the irrigation system as well as provide an essential nutrient for the plants. Field tests have proven that crop yields will increase substantially when potassium hypochlorite is the irrigation disinfectant.

5.23.5 Swimming pools

Swimming pools have used on-site generators for over 2 decades in both England and the United States. Large commercial and university pools are an excellent fit for the disinfection quality and product safety. Pool systems with either chlorine gas or commercial sodium hypochlorite take advantage of this product when consideration is given the number of people potentially exposed to material leaks. Many systems combine ozone and chlorination for more effective control of the chloramines and pool clarity.

Today pool disinfection control systems generally utilize an oxidation reduction potential (ORP) in combination with hand-tested chlorine results to assure that the pool chemistry remains safe and stable.

5.23.6 Cyanide destruction

The authors know of only one system installed for cyanide destruction purposes. This system was a small brine system used to destroy the cyanide in a wire manufacturer's waste stream. While this is an effective application of on-site systems the application opportunities are very few.

5.23.7 Industrial bleaching

Industrial bleaching using on-site hypochlorite systems began in 1976. This equipment was used primarily in large commercial and hospital laundries where there is an excellent fit for systems of 20 to 50 lb/d. These systems were generally used to their demise and seldom replaced.

5.23.8 Odor control

Sodium hypochlorite added to an air scrubber tower provides reasonable control in odor control applications. Composting and rendering facilities are two applications familiar to the authors where brine electrolysis has been used. In each case the operation was satisfactory when the power supply and controls were designed as NEMA 4X equipment. Composting applications exposed the equipment to high humidity causing corrosion of unsealed equipment. Rendering plants have the unique opportunity to demonstrate the effect of airborne fat. Everything becomes coated in a gummy mix causing equipment without proper protection to fail, thus the need for NEMA 4X equipment. Where there is an isolate and air conditioned room you may use equipment that does not meet the NEMA 4X standard, otherwise you will suffer failures if you do not adhere to this doctrine.

5.24 Seawater system applications

5.24.1 Power stations

Power station circulating cooling was the first major use of sodium hypochlorite generation with seawater as the cell feed stock. Engelhard Minerals used the Chloromat tubular cell followed one to two years later by DeNora and Diamond Shamrock. For any station located on a saline body of water this form of circulating water system treatment is an ideal electrolysis application. The station produces power and the salt for electrolysis is available for the cost of pumping power. System designs will allow you to use nearly any salinity level because the equipment can be designed to perform in the most severe conditions.

5.24.2 Oil field water

The first major electrolyzer system for oil field water flood systems was provided in the mid-1970s to Saudi Aramco for the Ghawar oil field. The water treatment plant for this field located on the Arabian Gulf was designed to process 4.5 million barrels of seawater and required a disinfection system of 12,000 lb/d. This system has been operating continuously since its installation and expanded by another 2400 lb. Systems of much smaller scope have been installed on platforms and floating production facilities worldwide.

5.24.3 Cooling system applications

Industrial cooling water applications include refineries, fertilizer plants, LNG plants, chemical processing facilities, steel mills, and etc. With minor variations all of these facilities have the same disinfection requirements as power stations, therefore one would apply the same design principles to equipment for these facilities.

5.24.4 Oil platforms

Oil field platform cooling and firewater disinfection applications can be considered the same as those noted above. The major variations on the design theme occur when the environmental conditions are considered. Here you must pay heed to the area classification and the salt air conditions for all your equipment designs. As oil becomes more difficult to locate, more offshore fields may be developed requiring this equipment.

5.24.5 Mammal pools

A rather unique application for seawater electrochlorination is the treatment of mammal pools housing dolphins and whales. This equipment is used in concert with ozone for bacterial and chloramine control. Electrochlorinatin systems have been used for many years at such noteworthy facilities as the Mirage Hotel in Las Vegas with excellent success.

chapter 6

Electrolysis system design considerations

In this chapter, the authors will highlight issues and problems to approach with caution when designing electrochlorination systems. The authors will not attempt to give code recommendations. Nationally accepted design codes should be adhered to very scrupulously. In addition to nationally accepted design codes, local design codes must be reviewed and compared with the national codes to confirm compliance across all code requirements.

Codes that encompass the electrical, mechanical, and environmental areas are pertinent to system design regardless of the equipment installation location. For equipment supplied outside the United States this is doubly true. With few exceptions, electrical code requirements are the most important to apply, followed by environmental codes. United States Environmental Protection Agency, National Electrical Manufacturing Association, and the American Society for Testing of Materials codes will provide equipment relatively violation-proof in the United States, while each country has its own set of similar regulations, as painful past experience has proven.

Beyond the environmental codes there is the environment itself. Here we cannot caution the reader enough to look at the operating environment with a very jaundiced eye or severe costly warranty headaches can soon follow. Corrosion caused by humidity and salt air will create electrical problems in tiny crevices that cannot be seen or foreseen until they occur. The problem with looking carefully at this issue is that the system cost is escalated at the quotation stage of the process by additional or more costly equipment.

As a supplier, more expensive equipment or additional equipment is not an easy scenario when one is in a competitive bidding situation. The balance between equipment reliability and cost is often difficult to recognize or maintain. The bidder must balance losing the project with premature equipment failure. The cost is clear when one considers losing the project, not so with a warranty issue. When a project is overseas, even minor warranty problems become major expenses where the cost of the equipment is usually much less than the internal material handling and shipment to the site.

6.1 Brine hypochlorite systems

Brine hypochlorite systems are typically used for disinfection of drinking water (wells or surface water), wastewater disinfection, cooling water applications, or for food processing operations. With very few exceptions, most of these sites are indoor installations where climate control is not a design issue. If one must install a system in the elements, then serious freeze protection is essential for the cells, water lines, and product storage tanks. Sodium brine solution will not freeze until below –20°F (–29°C) and yet it is prudent to heat trace and insulate the lower section of these tanks.

Low water temperature or pressure can be a problem for brine hypochlorite systems. Sites with feed water colder than 60°F (15°C) may require one to warm the water to help prevent premature electrode coating failure. Another problem caused by water colder than 40°F (5°C) is a reduction in cell performance efficiency. Here again one must increase the feed water temperature to maximize performance. Low water pressure requires the installation of a booster pump in the water feed system. Pressures below 40 psig are not normally acceptable for reliable brine hypochlorite system operation.

To get the most efficient operation from on-site electrolyzer cells, the cell feed temperature and the temperature rise through the cell are best controlled between 60°F (15°C) and 95°F (35°C). Having maintained the feed water temperature and cell temperature rise in this range, one should also reduce the sodium chlorate in the product caused by chemical inefficiency reactions and localized electrode reactions. It may be difficult to control inlet water temperature on small remote sites. Therefore, it is usually applied to sites of over 50 lb/d.

Larger systems usually include tanks of several thousand gallons (greater than 10 m^3) in the hypochlorite generation system. Excessive sun and increased temperature in the product hypochlorite storage tanks will cause accelerated degradation due to increased temperature and UV-light exposure. Fiberglass tanks pose a less severe degradation problem than translucent tanks because fiberglass tanks are protected by an exterior white gel coat. Also, proper anchoring of larger tanks is essential to account for high wind loads.

Available AC power at the site is important to system design to ensure the correct DC power supply. Wells and small system sites will most often have single-phase power, larger sites three-phase power. In any of these situations, the control panel will operate from 110 V or less using a step down transformer to achieve the lower voltage. In all cases, the power service at each installation must be sufficient to assure that the AC feed line current is acceptable and meets all codes when all the equipment is in operation.

System sizing is normally based on a system's daily demand for disinfectant at peak flow conditions. Equipment requirements must then be defined by the necessary dosing residual, the desired excess on-site generator production capacity, whether redundant generator equipment is important, the excess storage capacity to meet Class 2 reliability, and the dosing control scheme employed.

Having defined the production requirements, equipment sized for 100% of necessary capacity is advisable. A more suitable size would be 50 to 70% of a generator's rated capacity. If the sodium hypochlorite production requirement is very large, it is advisable to put in multiple systems of smaller capacity, for example, three units of 50% generation capacity. This design allows the performance of maintenance without loss of production capacity.

Redundant tanks are the accepted standard for on-site hypochlorite generation installations. These tanks provide backup storage and allow for bulk commercial hypochlorite storage in the rare event that the on-site system is inoperable for any length of time. The installation should include a mixing system to enable commercial hypochlorite dilution to a 1% solution for proper dosing control. Also, the diluted commercial hypochlorite solution will have a much lower rate of degradation than commercial sodium hypochlorite.

Dosing control schemes are available in three basic formats, manually adjusted, flow paced, and residual paced. There are instances where it is desirable to combine flow and residual pacing. This combination is accomplished using a programmable logic control (PLC) system. Well sites will most often have a manually adjusted dosing system since the well pump flow is constant and the hypochlorite product has a stable concentration. Flow and residual pacing require an analog signal be sent to the control panel, in turn signaling the variable speed drive on the dosing pump to adjust the flow to the proper residual. Part of the PLC function is to control the disinfection dose rate of change to prevent severe oscillations in the concentration of disinfectant applied.

An average dose of one to five milligrams per liter of sodium hypochlorite is generally required for potable water. Wastewater dosing may vary up to ten milligrams per liter. Isolated cases exist where the potable water dose required for a reasonable residual can be much higher. Two examples of this increased dosage requirement are for iron and manganese control and to treat for high ammonia levels. For example, potable water doses up to 40 mg/l of sodium hypochlorite may be required to remove ammonia through breakpoint chlorination.

Dosing volume and pressure will define the pump type required. Small systems will generally use a diaphragm pump for sodium hypochlorite application. Systems greater than 100 lb/d will often use positive displacement hose pumps capable of higher volumes. PVDF, Teflon, Viton, titanium, Hastelloy C, and ceramic are the construction materials of choice for pumps. Money extended on these materials will serve as an insurance policy to assure reliable system operation.

6.2 Seawater hypochlorite systems

Seawater electrolysis installations (onshore or offshore) are used where a ready supply of saline water sufficient to electrolyze into disinfectant sodium hypochlorite exists. Generally speaking, if there is a modest amount of salt, greater than 2500 mg/l, electrolysis can be performed using a seawater cell.

The applications outlined in Chapter 5 listed some of the design issues associated with seawater sodium hypochlorite systems. These issues are discussed in detail below.

6.2.1 Local regulations and codes

Local regulation and code review is much more important for seawater systems than for brine systems due to their broad application worldwide. Electrical codes can be particularly onerous when care is not taken in the review process. Area classification for offshore systems, whether by Europe or the United States, define specific needs to meet certain system limitations. Although mechanical codes are equal whatever the origin, review is just as important as with electrical code review.

Water discharge limitations are a consideration where local environmental regulations require stringent discharge controls. Most assuredly, this issue will change the treatment approach employed at a specific site.

There have been instances where the site discharge requirements caused severe fouling of the seawater cell. Although the generation system was capable of producing more sodium hypochlorite, it was constrained by discharge limits. There have also been cases where a constant sodium hypochlorite concentration could not meet the requirements even when no dosing limitations were enacted. For example, seawater clams and oysters $1/4$ inch to 2 inches in diameter grow very efficiently in a 0.3 to 0.4 mg/l environment with high flow rates, warm water, dark, no predators, and with high nutrient levels passing through the filters. Put another way, manually removing clam shells from electrolyzer cells each day for 2 to 3 weeks in Saudi Arabia's 110° heat is no picnic. A power station located in Indonesia grew very small snails in the circulating system when the chlorination level was less than 0.2 mg/l.

6.2.2 Site type

Site location and environmental conditions may control the approach taken to provide reliable equipment. Whether the electrolyzer site is onshore or offshore, the atmosphere will be salt laden. As well, desert locations will most assuredly have very high densities of dust and dirt during sand storm conditions. With a few exceptions, large systems, over 1000 lb/d (450 kg/d) are installed outdoors and protected from the sun. Offshore systems, although quite small, are seldom indoors and therefore exposed to whatever elements the site offers.

6.2.3 Local environmental conditions

Ambient air temperature and humidity should characterize material designs for the DC power supply and the control panel; the cooling method and media are important considerations to ensure reliable operation. Equipment exposure to temperatures well below zero is not advisable without proper freeze protection. On the other hand high temperatures will damage equipment if cooling methods such as shading and air conditioning are not employed. The authors experienced temperatures exceeding 180°F (82°C) in control panels installed in Saudi Arabia unprotected from the sun.

Water temperatures are important to consider in seawater sodium hypochlorite systems because the anode coatings are sensitive to temperatures below 60°F (15°C). In cold water, the electrode coating is affected by the increased oxygen concentration in the water. This increased oxygen concentration results in increased oxygen generation by the anode. To overcome this problem, several methods may be employed (e.g., reduction in the operating current density and changes to the electrode coating to allow the use of higher oxygen concentrations in the water).

Pipe, valves, and fittings must be designed for severe sun exposure (UV rays) and ozone conditions in rare occasions. PVC pipe will usually surface harden under these conditions and result in embrittlement, normally not a problem unless the pipe receives a severe impact. The effects of both UV and ozone on PVC pipe may be controlled by painting the pipe. Nearly any latex paint, preferably a light color, will serve to block UV and ozone effectively. With weather exposure, FRP pipe surfaces may expose glass fibers. Painting FRP pipe may also solve these problems. However, specific painting steps may be necessary for effective painting. These steps may include the use of specific wash coat primers to ensure good paint adhesion.

FRP tanks should have a white gel coat as provided by the tank manufacturer to prevent gradual surface degradation.

Steel fabrications should always be seal welded to eliminate crevice corrosion problems. The long-term corrosion control benefits may justify the minor added cost of seal welding. Painting fabricated steel requires coatings having at least a three-part system. This system provides a zinc base coat to provide severe duty protection. Properly prepared and painted steel will withstand very harsh salt environments and random hypochlorite exposure.

6.2.4 Available water pressure

System water supplies are determined by the source water pressure. Offshore systems seldom require booster pumps because their cooling water systems have sufficient pressure to support electrochlorinator requirements. Onshore systems nearly always require seawater booster pumps where the circulating cooling systems use low-pressure high volume equipment. The supply pressures most often used in electrochlorinator systems are in the range of 100 to 150 ft (47.2 to 70.8 m) of discharge head.

6.2.5 Dosing point pressure requirements

Dosing point pressure generally is low for circulating pump intake structures and for offshore platform pump caissons. These systems only require seawater booster pumps to provide both the seawater supply and the required pressure to get to the dosing point.

Systems having high dose point pressure requirements usually require a degassing tank and independent dosing pumps in addition to the seawater booster pumps. The dosing tank serves as a positive suction head pump supply and will usually require air fans to dilute the by-product hydrogen. A few plants have used open top tanks for free hydrogen escape without the need for fans. These systems are very effective and reduce required electrochorination system equipment. A note of caution: When one is working with deep water hypochlorite injection situations, one must account for the head losses below seawater level.

Be aware that seawater electrolysis produces magnesium hydroxide and related calcium waste products. These deposits will settle in the storage tank unless the tank and piping is designed to keep them in suspension. While not all of these deposits can be removed, keeping them in suspension allows them to be pumped from the tank.

6.2.6 Dosage control methods

Pump configurations for dosing systems most often used are: a single pump in continuous dose operation, a single pump in shock dose operation, and a single pump for continuous operation using a second continuous operation capable pump for shock dosing purposes. In all cases, regardless of the site, each pump in operation has a second pump in position to support failure occurrences.

6.2.7 Power supply requirements

Three-phase supply power varies from more common 380, 415, and 460 V to 4,160, 6,600, and 11,000 V available. Systems having large DC power units will frequently use 4,160 or greater voltage for large onshore plants. Designers often prefer to use high voltage power, when available, to allow the use of smaller sized cabling resulting from lower AC operating line currents and existing high voltage switch gear. Depending upon the plant power distribution system, control voltage power is stepped down from the higher source power by a transformer to 110 or 220 V from the main power supply line. Power also may be supplied separately from an independent source. Systems installed offshore, with few exceptions, use supply power of less than 600 V.

6.2.8 System sizing

System sizing is determined using several basic criteria: seawater chlorine demand, dosing method (continuous only, shock only, combined continuous and shock), seawater salinity, and temperature.

Chlorine demand is quite variable yet generally within a narrow range of 2 to 3 mg/l. For example, in the Arabian Gulf the seawater chlorine demand has been tested at 2.7 mg/l and in Singapore it was tested at 2.4

mg/l. There are always exceptions to this general rule. However, those exceptions are rare and usually do not result in undersized systems.

Dosing method is a matter of past engineering practice and the influence of previous chlorine gas disinfection operations on the dosing system design. Many retrofitted electrochlorination systems will have both continuous and shock dosing capabilities. These capabilities are similar to those used in gas chlorination systems. In these dual systems, continuous dosing rates of 1 to 2 mg/l are the most common. Shock dose rates can range from 3 to 10 mg/l for periods of 15 to 60 min depending upon the conditions. The dosing regimen is integrated in the control panel operations to assure cooling system performance is maintained.

Based upon the chlorine demand data, discussed above, systems are initially sized to provide a continuous dose rate of 1 to 2 mg/l. This practice of sizing electrolyzers to provide 1 to 2 mg is commonly employed for most seawater electrochlorination systems, unless the client specifies otherwise. The cooling systems in offshore sites function well when treated in this manner for two reasons. First, the small larvae naturally occurring in the seawater do not like the irritation associated with low levels of chlorine. Secondly, bromine (approximately 50 mg/l) naturally present in seawater is converted to hypobromous acid during electrochlorination. This acid is a very effective biocide at the normal seawater pH of 8.5 where the hypochlorous acid is less effective. Thus, the presence of bromine helps to improve the efficacy of seawater generated hypochlorite when compared to chlorine gas alone.

All dosing can be stopped for brief periods if an equipment failure occurs. However, it is not prudent to be offline for longer than 24 h. After that period the larvae for barnacles and bivalves will begin to deposit on the circulating cooling system's walls and crevices.

As discussed earlier, electrolyzer equipment capacity may be sensitive to low seawater temperatures and low salt content conditions. Seawater salinity of 80% (as chloride) of nominal seawater chloride content (18,980 mg/l) and temperatures greater than 60°F (15°C) are required for normal electrolyzer operation. In these cases where salinity or temperature is less than these values, additional electrolyzers must be used to reduce the operating current density in the cell. Alternatively, the electrolyzer electrode coatings may be modified for operation at the lower temperatures and salinity.

The addition of electrolyzers to a given system is accomplished easily, although the designer should be aware of the cost of the added electrolyzers. The multiple cell modules design concept may provide for lower operating current and operational redundancy. Designing excess capacity into large onshore systems is based upon the chlorination requirements discussed above. It is common to design with 2 × 50% modules, 3 × 50% modules, or 3 × 35% modules to provide excess capacity and redundancy.

Another consideration in system sizing is the possibility of future plant expansion. In these cases, one may wish to consider the addition of seawater supply pump capacity and piping as well as other applicable civil and electrical requirements.

6.2.9 Biofouling control

Biofouling control is managed by varying the electrolyzer operating current with manual or automatic controls to maintain a desired chlorine residual in the circulating stream of the electrolyzer system.

Manual control is the most frequently employed method of maintaining an appropriate chlorine residual in the circulating stream. Plant laboratories or operations personnel will perform a manual chlorine residual test, on a regular basis, daily or weekly, and the system output is adjusted via the rectifier operating current. Experience has proven that weekly tests are usually sufficient to provide effective residual control due to the slow seasonal change in seawater chlorine demand.

Automated control is normally achieved by using a chlorine residual analyzer. This instrument sends an analog 4 to 20 mA signal to a controller or the PLC control system with a control scheme that in turn provides an analog signal to the DC rectifier for current control. The control loop is essential to prevent the DC rectifier current oscillation caused by control sensitivity. As noted earlier, the system will be very stable because, with the exception of a plankton bloom, seawater chlorine demand changes are usually very slow.

chapter 7

Economic evaluation principles for electrolysis systems

7.1 System installation

The economics of site engineering begin with site preparation including concrete pads, shelter or building installation, and required water and power services.

The authors have been responsible for or participated in the installation of electrochlorination (EC) plants worldwide sized from 20 lb (9 kg) per day to 57,600 lb (26 t) per day. Each site provided an opportunity to oversee the requirements for both large and small sites.

A site's location is important for effective EC system cost evaluation. For example, seawater EC sites in Saudi Arabia require significant import cost for raw materials and labor to complete a plant. A similar plant in Taiwan or Singapore will have most of the required construction materials, i.e., rebar, concrete, lumber, etc., readily available locally. Local labor does not exist in countries such as Saudi Arabia, further adding to the contracting cost. In such a case, labor personnel are imported from Pakistan, Sri Lanka, Philippines, Indonesia, etc. at low wages and contracts of 2 to 3 years.

Installation time for a plant can vary widely for reasons of site location and equipment scope. A 1000 lb/d (450 kg/d) domestic brine electrolysis site having an existing building will usually require 6 to 8 weeks to install and commission. This time allocation assumes that no civil work is required to complete the installation. If civil work is not completed one must allow 12 to 16 weeks to complete installation and commissioning. Installation and commissioning will require from 6 to 12 weeks of supervisory time after all civil requirements are completed.

An undeveloped site is another matter entirely. This discussion will provide an overview of the necessary elements to complete an installation.

There are many small installation and commissioning issues that require one to constantly monitor contractor progress. Dependence on the local contractors to complete their portion of the installation demands careful coordination and good cooperation.

When evaluating the site cost an engineering firm considers electrochlorination plants as an offsite piece of plant. That is to say, it is not essential to main plant function. As such, this equipment is added as an incremental part of the main installation and the contractor treats main plant installation issues with more urgency.

Overall installation costs for an on-site generator as described above will be from 35 to 100% of the generator equipment cost. The complexity of site requirements define these costs; e.g., does the system require a building, is there air handling equipment required, is the installation an expansion, etc. Therefore, unless a better cost basis is available, one may consider installation cost to be approximately 50% of the EC plant equipment capital cost.

7.2 System operations

7.2.1 Operating economics

The major cost factors in a complete evaluation of on-site sodium hypochlorite production are direct costs for water, salt, power, operating and maintenance labor, and operating supplies; fixed costs for fixed charges, deprecation, taxes, and insurance; plant overhead; and general expenses of administration and financing costs. The following narrative defines the most important issues when evaluating on-site generation operating costs. It is not the intention of the authors to provide a complete analysis in this text but to provide a cost basis for the readers to begin their analysis.

Brine systems require salt and power to supply the requisite sodium hypochlorite. The same is not true of a seawater system since salt is freely available. This difference has a significant impact on disinfectant cost.

For the purposes of this discussion, water does not contribute to on-site production cost since water used in the generation process is returned to the distribution system as part of the disinfectant.

7.2.1.1 Brine system operating costs

Power cost for on-site hypochlorite production, while widely variable, contributes more than 50% to the production cost at 2.5 AC kWh/lb (5.5 AC kWh/kg). Most plants cannot see the power usage because the system is so small compared to other plant power consumers. For example, a 50-horsepower pump motor will consume 37 kW and a 5-lb/h (2.25 kg/h) hypochlorite generator will consume 12.5 kW. Most plants operating this size system also operate many large power consumers. Therefore, the hypochlorite generation cost is somewhat hidden or masked in the day-to-day power consumption variations. The power cost applied in this discussion is $0.10 per kWh.

chapter 7: Economic evaluation principles for electrolysis systems

The need for separate salt delivery contracts to the plant site does not allow the same innocuous condition as seen with power usage. The need for a brine system salt supply requires a readily available source of good quality salt at reasonable prices near the electrolyzer installation. These regular deliveries generally require long-term contracts for a reasonable cost basis.

Three major categories of salt are sold in bulk, mined salt, solar salt (evaporated sea salt), and purified salt. The cost of mined and solar salt is considered equal, while purified (table grade) salts may (depending upon your geographic location) cost 1 to 3 times that of mined or solar salt. Since brine systems presently on the market do not require purified salt for reliable operation, mined and solar salt will be the focus of the discussion.

The accepted standard for salt usage in brine systems is 3.5 lb salt per pound of chlorine generated as sodium hypochlorite. Salt prices will fit into a range of 0.02 dollars/lb (0.045 dollars/kg) to 0.10 dollars/lb (0.22 dollars/kg) dependent upon quantity, quality, and transportation requirements. The standard cost used for cost analysis purposes is 0.05 dollars/lb (0.11 dollars/kg) delivered to the site.

The remaining items in a direct cost analysis are operating and maintenance labor requirements, operating supplies, and a chemical index inflation allowance. Labor is required for system oversight and to perform any required maintenance for cleaning or equipment repairs. Operating supplies refer to maintenance materials, acid for cleaning, etc., and replacement electrodes whose reasonable life is 5 to 7 years. The inflation index uses chemical price indices for material price inflation.

Labor of 15 minutes per day is the allowance for the two most important maintenance requirements, cell cleaning and electrode replacement. The labor cost used at the time of this writing is $12.00/h excluding plant overhead.

Without regard for EC plant warranties and because the authors have no exact electrode replacement charges, a charge of $0.02/lb (0.045 dollars/kg) of chlorine produced will be assumed. This accrual figure should provide replacement electrodes after 5 years of system operation between replacements.

The chemical price index used is the M&E index. The historical price basis is established and offers adequate accuracy for cost analysis. In this discussion, an inflation number of 3% based upon the consumer price index will be used. This number is high for the present, however it is a long-term average that will prevent gross analytical errors.

In summary, for the cost of product at 100 lb/d (45 kg/d) with no water charge, salt at $0.175/ lb of Cl2 ($.05 x 3.5 lb), power at $0.25/lb of Cl2 ($0.10 x 2.5 kWh), unburdened labor at $.02/ lb of Cl2 ($12.00/60 x 10 min/d), the consumer price index inflation at 3% of the total fixed costs equaling $.015/lb of Cl2, the chemical price index inflation at 5% of the total fixed costs equaling $0.022/lb of Cl2, and an electrode replacement cost of $0.02/lb of chlorine. The total price of chlorine produced with a brine type sodium hypochlorite generator on-site is $0.502/lb of chlorine.

Because the general labor and index cost structure is nearly equal regardless of the site location within a country, we tend to view production cost as simply the cost of salt and power as in the example above where the raw material cost is $0.425/lb of chlorine ($0.175 plus $0.25).

The most recent brine electrolyzer cell developments will further reduce system salt and power consumption. The authors expect a reduction in cost which will approach the cost of on-site production of small quantities of chlorine gas.

7.2.1.2 Seawater system operating costs

Seawater systems eliminate the requirement for salt cost. The above brine system analysis provides a foundation for the reader to understand seawater electrolysis costs.

To elaborate further, the authors will adjust the summarized cost above. Bearing in mind that the seawater electrolyzer cells are more power efficient than brine electrolyzer cells, there is an inherent power cost advantage. Seawater system costs 1.8 kWh versus brine system product at 2.5 kWh per pound of chlorine.

To discuss seawater system cost analysis, a production requirement of 1000 lb/d (454 kg/d) chlorine equivalent requiring seawater booster pumps and no product dosing pumps will be used. The cost of power for the seawater booster pumps is assumed to be 5 kW for increasing the supply pressure only. Booster pump power consumption is 5 kW times 24 h equaling 120 kWh divided by 1000 lb/d chlorine production equaling a pumping cost of $0.012/lb of chlorine.

Assuming the plant must buy power at $0.1/kWh, production power cost is $0.192 lb of Cl2 ($.1 × 1.8 kWh plus $0.012), unburdened labor at $.02/lb of Cl2 ($12.00/60 × 10 min/d), the consumer price index inflation at 3% of the total fixed costs equaling $.006/lb of Cl2, the chemical price index inflation at 5% of the total fixed costs equaling $0.011/lb of Cl2, and accrual for electrode replacement of $0.02/lb of chlorine. The total price of chlorine produced with an on-site seawater sodium hypochlorite generator is $0.249/lb of chlorine.

chapter 8

Electrolysis system installation, operation, and maintenance

The intention of this chapter is to provide the reader with practical electrochlorination system installation, operation, and maintenance requirements. These requirements are for reference only and in no way are intended to usurp supplier recommendations. While brine and seawater electrochlorination systems are very similar they are separated in this chapter to avoid confusion. Note, no consideration is given in this discussion to civil installation requirements.

The lists below are general requirements to install an on-site hypochlorite generation system, provide estimates of equipment and manpower, and the approximate time to install the system size noted. One will then have a reasonable order of magnitude to judge the requirements for larger or smaller system installation requirements.

Mechanical and electrical installation requirements, hydraulic pressure testing, and commissioning time requirements for either a brine system or a seawater system of 1200 lb chlorine equivalent per day are shown in Tables 8.1, 8.2, and 8.3, respectively.

Outlined below are the major checkpoints for the start up and operation of a brine electrolyzer system:

1. Confirm that all anchor locations are approximately correct.
    a. Note: Equipment installation using permanently located anchors is not recommended. Anchor tabs or concrete block-out holes in the foundation are preferred for installation ease, leveling requirements, and the necessity to grout equipment regardless of pad condition.
    b. When the block-out method has been employed, use a zero datum reference point as a location basis, then use a chalk line to correctly locate the equipment.

Table 8.1 Equipment Weight

Equipment Qty	Description	Estimated Dry Weight (lb)
1	Cell system	1800
1	DC Rectifier	1200
1	System control panel	500
1	40 ton dissolver	1500
2	20,000 gallon tanks	1500
2	Dosing pumps	150

Table 8.2 Equipment Installation Requirements

Installation	Installation Duration	Manpower
Set equipment	1 week	6 people + 1 supervisor
Pipe installation	2–3 weeks	3 people + 1 supervisor
Install Electrical	2–3 weeks	3 people + 1 supervisor

Note: Equipment setting requirements assume use of a 20-ton truck crane and a front end loading fork truck

Table 8.3 Equipment Testing Requirements

Testing	Test Duration	Manpower
Hydro test	3–5 days	2 people + 1 supervisor
Electrical test	3–5 days	2 people + 1 supervisor
Commissioning tests	3–5 days	1 person + 1 supervisor
Performance tests	1–3 days	1 commissioning engineer

2. Set all major equipment in their respective positions.
3. Adjust all equipment to level and grout in position.
4. Set anchors or anchor tabs and grout in or tighten in position.
5. Using a chalk line, lay out piping and conduit positioning.
6. Position and anchor pipe and conduit supports.
7. Install pipe and conduit on their respective supports in accordance with site drawings.
8. Complete flanged and threaded connections for all pipe and conduit.
9. Pull all power and electrical wire to the designated termination boxes.
10. Some sites will require a megger test of all installed wire. In this case you must disconnect the terminations and test to assure there are no voltage leaks to ground.
11. Perform termination wring-out tests on all power and instrument cabling.
12. Terminate power and instrument wire.
13. Flood and hydraulically test piping sections starting at the inlet of the system. Progress through each pipe section until complete, repairing any leaks before moving to the next test section.

chapter 8: *Electrolysis system installation, operation, and maintenance* 89

Next is the pre-commissioning checklist:

1. Piping connections
   a. Confirm that all piping is installed and properly connected.
   b. Confirm that all threaded and flanged connections are tight.
   c. Hydrostatically test all piping to site specified pressures or to 1.5 times the system's nominal operating pressure.
2. Power connections
   a. Confirm that all electrical equipment meets local, state and federal codes.
   b. Confirm that the motor control center is installed and connected properly and all electrical terminations are tight.
   c. Disconnect and meager test the main power supply wiring from the power source to the motor control center and from the motor control center to the DC power supply to assure there is no damaged wire or connection ground faults.
3. Instrumentation wiring
   a. Perform loop checks of all instrumentation wiring to confirm all instrument terminations are properly connected.
   b. To assure that all analog signal systems are functioning properly, use a digital signal generator to confirm proper field instrument operation.
   c. Confirm that all terminations are tight at the control panel and the field instruments.
   d. Apply anti-fungal lacquer to all field and control connections, if required. This should be completed only after the systems are thoroughly tested to assure no further changes are required.
4. Pump testing
   a. Remove the flexible coupling guards and disconnect the flexible couplings.
   b. From the motor control center, bump the motor starter assembly to confirm proper rotation. Correct motor rotation as required.
   c. Reconnect the coupling assemblies and align the motor and pump coupling per the manufacturer's instructions.
   d. Reinstall the coupling guards.
   e. Remove the pump bearing oil addition plug and measure the bearing housing oil level to confirm that there is sufficient oil on the bearing housing. Making certain that you have the manufacturer recommended oil, add oil as necessary to the bearing assembly.
5. For pumps with seal flush systems
   a. Disconnect the seal flush supply lines and flush the lines at full flow for 15 min to remove any debris from the lines.
   b. Reconnect the water supply line and set the water flow at the proper flow rate. This flow for most seal assemblies is 0.5 gpm (2 l/min).

6. Inlet automated strainer testing
   a. Energize the strainer control panel.
   b. Open the outlet strainer isolation valve and supply water to the strainer by opening the inlet isolation valve sufficiently to fill the strainer. This is to assure that you have some water in the strainer to assure slight lubrication of the scraper assembly during testing.
   c. Start the manual strainer backwash sequence to confirm that the strainer motor is functioning. Stop the motor as soon as you have confirmed proper motor operation.
   d. Confirm that the strainer drain valve opens and closes when the system is in backwash.
   e. Open the pressure differential switch and short pressure switch control contacts to the strainer control panel. This should begin the pressure switch initiated strainer backwash sequence. Confirm that the sequence of operations is correct and that the duration of operation is per the expected local operating requirements.
7. Dilution fans
   a. Confirm that the power supply wiring is properly connected.
   b. Remove the fan drive belt covers.
   c. Remove the fan drive belts.
   d. From the motor control center, bump the motor starter assembly to confirm proper rotation. Correct motor rotation as required.
   e. Reinstall the fan drive belts and the drive belt covers.
   f. Start one of the fans and confirm that the in-line airflow switch is functioning properly.
8. Systems with water softener controls
   a. If the softener is an electrically operated unit, connect the softener valve to the control power.
   b. The water softener timer is normally factory set but should be tested and adjusted to local water hardness conditions. Instructions for setting the regeneration cycle should be in your operating manual.
9. Systems with storage tanks and level controls
   a. Manually fill the storage tank to 60 to 75% of capacity.
   b. Open the tank manual outlet isolation valve to the pump suction and the pump suction manual isolation valve at the pump.
   c. Close the pump discharge manual isolation valve.
   d. Confirm that the level control sensor, whether electronic or air, is functioning properly by varying the signal, again electronic or air, to the level control valve.
   e. Supply power or air to the automated level control valve.
   f. The storage tank level control valve should be open because the tank is partially filled and the level sensor is functional.
   g. Start the pump with the pump outlet valve closed.

h. With the pump in operation slowly open the pump outlet manual isolation valve. Observe that the automated level control valve closes as the tank level recedes to the level control set point.
i. Stop the pump and isolate the tank in preparation for system start up.

8.1 Brine systems: general commissioning procedure

The procedure is as follows:

1. Disconnect all electrical power from the DC rectifier and controls.
2. Disconnect the incoming water line to the water softener and flush the line at full water flow for at least 10 min, then stop the water flow and reconnect to the water softener.
3. Fill the salt dissolver to at least the 50% level.
4. Engage the power to the system control panel.
5. Confirm that proper power is available to the dissolver controls and level control valve.
6. Open the manual valve supplying water to the salt dissolver.
7. Allow the brine dissolver to fill to the proper level. Be certain to confirm that the level controls close the automatic water supply valve to stop water flow.
8. Allow the salt to dissolve for at least 4 h and preferably 12 to 24 h.
9. Open all manual cell outlet valves to the system storage tank.
10. Close the system flush and drain valves.
11. Close cell water and brine feed control valves.
12. Open the supply water valve to the water softener at the recommended pressure, usually greater than 40 psig, adjusting as required.
13. Slowly open the water supply valve to fill the electrolyzer cell and bleed all the air from the system piping to the storage tank, usually 3–5 min.
14. Set the water and brine flows at the recommended rates. Allow about 10 min to fill the system with brine solution.
15. Turn the DC rectifier control settings to 0.
16. Engage the power to the DC rectifier. Reconfirm that the AC power to the rectifier is at the proper voltage.
17. If your rectifier has a voltage control, increase the DC voltage control to 100%.
18. Start the rectifier power output by increasing the DC current control pot to 25% of rated output and hold at that current for 15 min.
19. Test the incoming AC power supply lines to the rectifier to confirm that all three phases are operating with balanced current, within 5%.
20. Increase DC power to the cell to 100% and allow it to operate for 30 min. Once again confirm that the three power supply phases have balanced current.

21. Using a DC voltmeter, test the cell operating voltages after the system has operated for several hours at the nominal operating conditions.
22. Confirm that the remainder of the system is operating smoothly by following the required inspection and test protocols provided with your system.
23. Fill the storage tank to 30% of capacity.
24. Open the tank to pump supply valves.
25. Provide power to the system dosing pump and controls and manually confirm that the pump is operating properly.
26. Should you have an automatic dosing control system you must confirm that the pump control signal requirements are available and the span required for proper control is set in accordance to instrument instructions.
27. Switch the pump control from manual to automatic and confirm that the pumping system is functioning to design requirements.

8.2 Seawater systems: general commissioning procedure

The procedure is as follows:

1. Disconnect all electrical power from the DC rectifier and controls.
2. Disconnect the incoming seawater supply line to the seawater strainer and flush the line at full water flow for at least 10 min, then stop the water flow and reconnect to the seawater strainer.
3. Engage the power to the system control panel.
4. Confirm that the seawater strainer has proper voltage to the control panel and the strainer cleaning motor.
5. Engage the AC power to the seawater booster pumps and the product dosing pumps.
6. Confirm that the strainer drain valve is closed.
7. Close the seawater booster pump inlet and outlet isolation valves.
8. Open the strainer inlet and outlet isolation valves.
9. Open all cell outlet manual valves to the system storage tank.
10. Note: For a system with only a degassing device open the valves to the system dosing point.
11. Close the cell system flush and drain valves.
12. Close the storage tank outlet isolation valve.
13. Open the seawater booster pump inlet isolation valve.
14. Start the seawater booster pump manually from the system control panel.
15. Slowly open the seawater booster pump outlet isolation valve.
16. Confirm that the seawater pressure is at the recommended level.
17. Slowly open the seawater strainer inlet and outlet isolation valves. This will provide seawater to the cell module inlet isolation valve.
18. Note: This step must be completed slowly to prevent cell and PVC pipe waterhammer damage. Slowly open the seawater supply valve

chapter 8: Electrolysis system installation, operation, and maintenance

to fill the electrolyzer cell module at 20 to 25 gal/min and bleed the air from the system piping to the storage tank or degassing device, usually less than 5 min.
19. Set the seawater flow to the vendor recommended rate.
20. Turn the DC rectifier control settings to 0.
21. Engage the power to the DC rectifier. Reconfirm that the AC power to the rectifier is at the proper voltage.
22. If your rectifier has a voltage control device, increase the DC voltage control pot to 100% and turn the DC current control pot at 0.
23. Start the rectifier power output by increasing the DC current control pot to 25% of rated output and hold at that current for 5 min.
24. Test the incoming AC power supply lines to the rectifier to confirm that all three phases are operating with balanced AC current, within 5%.
25. Increase DC power to the cell module to 100% and allow it to operate for sufficient time to confirm proper operation.
26. Once again confirm that the three power supply phases have balanced current.
27. Confirm that the tank level control system is operational.
28. Confirm that the level control valve has power.
29. Using a DC voltmeter, test each cell's operating voltage in the cell module after the cell system has operated for several hours at the nominal operating conditions.
30. Confirm that the remainder of the system is operating smoothly by following the required inspection and test protocols provided by the manufacturer for your system.
31. If your system has a storage/degas tank, fill the storage tank to 60 to 70% of capacity.
32. Close the dosing pump discharge isolation valves.
33. Open the storage tank outlet isolation valve from the storage/degas tank to the dosing pumps.
34. Open the dosing pump inlet isolation valves.
35. When the dosing pumps are flooded start the dosing pump from the control panel and slowly open the discharge isolation valve.
36. Should you have an automatic tank level or dosing control system you must confirm that the dosing pump or tank level control signal requirements are available and the signal span required for proper control are set in accordance with the instrument instructions.
37. Switch the pump control from manual to automatic and confirm that the dosing or level control system is functioning to design requirements.

chapter 9

System design and trouble analysis

Systems designed today having the latest materials of construction will operate reliably for many years with a minimum of equipment maintenance. The most important maintenance is cell cleaning. This maintenance should be considered an essential part of a preventive maintenance program. All other equipment will require mechanical, instrument, and electrical maintenance.

When systems malfunction, identifying the problem may be difficult where the equipment environment is severe. The trend toward more complex equipment control schemes provides an environment for complex instrument and control failures. Offshore environments, Middle Eastern, and Southeast Asia environments must be designed for high humidity and heat. If not, the equipment may fail due to electrical junction corrosion faults. The use of antifungal lacquer on electrical terminations and controls after all equipment is in operation will assist in preventing these types of field problems.

Tables 9.1 and 9.2 provide some of the more common elements of electrochlorinator fault analysis. A brief discussion of fault analysis is provided as an example of an approach used to trace field problems. A high DC cell voltage problem is used in the following example.

The symptom for this example problem is excessive cell voltage. Due to this excessive voltage the operator is unable to maintain the operating amperage of the cell at the desired level. This low operating amperage is indicated by the rectifier voltmeter on the rectifier unit. Listed below are a series of checks and tests that would be used to identify the problem. Note, the example is divided into two procedures (one for seawater systems and one for brine systems).

Seawater Systems— Test the incoming seawater for salt content:

1. Excessive rain will reduce the salt strength and cause this high DC cell voltage problem.
2. The water temperature may have decreased to a temperature low enough to cause excess voltage through increased solution resistance.

Table 9.1 Symptoms – Seawater systems

Probable Cause	High temp shut down	High DC voltage	Low DC voltage	High current	Low or no current	No system power	Poor or no hypo production	Unit runs constantly	Flow loss
Low seawater flow rate	X			X	X		X		
DC current too high	X	X		X					
Cell deposit build-up	X	X					X		
Seawater flow high			X						
Bad DC electrical connection		X			X				
Current control not set correctly		X	X		X		X		
Water valves set improperly	X		X		X				X
Storage tank level control failure					X		X	X	
Cell flow switch failure							X		X
Temperature switch failure					X		X		
Control panel switches set incorrectly							X		X
Vent line clogged									
P.C. board failure				X	X	X	X		
Cell piping clogged							X		X
Cell electrode wear			X		X		X		
Loss of main AC power						X	X		X
Low AC supply voltage				X	X		X		
Control panel failure					X	X	X	X	X
Rectifier P.C. board out of phase			X		X		X		
Power SCR or diode failure			X		X		X		

This condition will not improve until the inlet water temperature improves.

   3. The cell has excessive calcarious deposit formation.
      a. Has the cell been acid cleaned as indicated in the maintenance manual?
      b. Measure the independent cell assemblies to confirm that no individual cell pack or assembly has been excessively damaged.
   4. The cell current is above the proper setting causing high cell voltage.
      a. The current control pot is set incorrectly.
      b. The control pot has failed.
      c. A control circuit board in the rectifier has failed.

Table 9.2 Symptoms – Brine Systems

Probable Cause	High temp system shut down	High DC cell voltage	Low cell voltage	High operating current	Low or no cell current	No system AC power	Poor hypo production or none at all	Unit runs constantly	Flow loss
Low flow rate — Water or Brine	X	X	X	X	X		X		
DC current too high	X	X		X					
Cell electrode deposit build-up	X	X					X		X
Salt brine flow too low	X	X			X		X		
Salt brine flow too high			X	X					
Bad DC electrical connection		X			X				
Current control not set correctly		X	X		X		X		
Improper valve settings	X			X	X	X			X
Storage tank level control failure					X		X	X	X
Cell float switch failure							X		X
Temperature switch failure					X		X		X
Panel switches positioned incorrectly					X			X	X
Vent line clogged									X
Rectifier P.C. board failure				X	X	X	X		
Cell piping clogged									X
Cell electrode wear			X		X		X		
Loss of main AC power					X	X			X
Low AC supply voltage			X		X				X
Relay or PLC failure				X	X		X	X	X
P.C. board out of phase			X		X		X		
Rectifier SCR or Diode failure			X		X		X		

5. There is a loose bus bar connection between the cells or between the cells and rectifier.
    a. This problem can occur occasionally due to bus bar expansion. The connection fasteners must be tightened.
6. The cell voltage is high and no other corrective actions improve cell voltage.
    a. The cell is near the end of its useful life. New electrodes will have to be installed to return the cell to its original condition.
        i. Do a cell voltage scan to determine which cell assemblies are causing a problem.

ii. Replace all of the electrodes in the cell that is the proven problem.

Brine systems— Test the incoming brine for salt content:

1. Brine systems have controlled water and brine flow, therefore check
    a. Water pressure loss
    b. Water flow loss
    c. Brine pressure loss
    d. Brine flow loss
2. Inlet water temperature has decreased.
    a. A water heating method must be provided either by using the product discharge heat or using a stand alone water heater.
3. The cell has excessive calcarious deposits.
    a. Has the cell been acid cleaned as indicated in the maintenance manual?
    b. Measure the independent cell assemblies to confirm that no individual cell pack or assembly has been excessively damaged.
4. The cell current is above the proper setting, causing high cell voltage.
    a. The current control pot is set incorrectly.
    b. The control pot has failed.
    c. A control circuit on the rectifier control board has failed.
5. There is a loose bus bar connection between the cell and rectifier.
    a. This problem can occasionally happen due to bus bar expansion. The connection fasteners must be tightened.
    b. The cell voltage is high and no other corrective actions improve cell voltage.
    c. The cell is near the end of its useful life. New electrodes will have to be installed to return the cell to its original condition.
        i. Do a cell voltage scan to determine which cell assemblies are causing a problem.
        ii. Replace all of the electrodes in the cell that is the proven problem.

chapter 10

System safety

Safety is an issue that must be considered whenever or wherever people are involved with equipment operation. Previous chapters have addressed the history, concepts, processes, specifications, operation, and maintenance of both on-site brine and seawater hypochlorite generation systems. Due to the importance of safety, this chapter provides the reader with a set of operating safety instructions applicable to any on-site hypochlorite generation system regardless of manufacturer or site.

Under normal operating conditions, hypochlorite systems are leak proof. However, leakage and spillage may occur during system operations. Listed below are safety instructions that can be used as guidelines to prevent injury to operating personnel.

Safety can be categorized in two areas: chemical and electrical.

10.1 Chemical safety

Two types of chemicals associated with on-site generation of sodium hypochlorite are considered hazardous and potentially dangerous to operating personnel:

1. NaOCl   sodium hypochlorite
2. HCl   hydrochloric acid

Be aware that the direct mixture of sodium hypochlorite solution with hydrochloric acid while making the cleaning transition process will release free chlorine gas.

Note: DO NOT MIX ACID WITH SODIUM HYPOCHLORITE.

10.1.1 Sodium hypochlorite handling (NaOCl)

Although the product from on-site sodium hypochlorite generation systems has a lower strength than commercially available sodium hypochlorite, the instructions below discuss the product as though it is the strongest hypochlorite available.

It is essential that safety precautions are taken when carrying out any duties which involve possible contact with sodium hypochlorite. Personnel must be instructed on the properties of sodium hypochlorite and local safety regulations for chemical handling must be adhered to at all times. First aid equipment must be readily at hand and a hosed water supply must be available before work is started. Eyewash bottles should be placed at strategic points throughout the hypochlorite plant.

The following information should be made available to all personnel:

- Sodium hypochlorite handling hazards
  - Ingestion of NaOCl by mouth has a very serious poisoning effect, which could lead to death.
  - Inhalation of NaOCl vapor can give rise to headaches, irritation of the mucous membrane, lack of coordination, and loss of consciousness.
  - Persistent absorption of NaOCl through the skin may lead to skin disorders.
- Splashes on the skin
  - Remove the affected clothing immediately and wash the skin thoroughly until all contamination has been removed.
  - Refer the patient to a doctor or a hospital for further treatment.
- Gassing and ingestion by mouth
  - Obtain medical assistance immediately.
  - Move the patient to a fresh air environment as soon as possible.
  - Keep the patient warm with blankets and resting quietly.
  - Oxygen must be administered by qualified personnel.
  - If breathing fails, administer artificial respiration immediately and continue until the patient breathes again or until a doctor instructs otherwise.
  - If NaOCl has been swallowed, do not induce vomiting but obtain medical aid immediately.

10.1.2 *Hydrochloric acid handling (HCl)*

Muriatic acid (a 30% solution of hydrochloric acid) is used for cell cleaning. This acid will cause serious chemical burns if it comes in contact with the skin for more than a few seconds. If the acid comes in contact with the eyes, it will rapidly cause serious damage resulting in impaired vision or total loss of eyesight. It is poisonous if ingested. Always wear eye protection, protective clothing, and rubber gloves when handling.

In order to minimize the risk of these hazards, certain precautions must be taken. These precautions are:

chapter 10: System safety

- Personnel must be fully instructed as to the characteristics of acid.
  - Always avoid contact with the eyes, skin and clothing.
  - Eye protection must always be worn when handling acid or during any operation where there is the possibility of acid leakage.
  - Certain protective clothing must also be worn. For normal tasks such as handling carboys, etc., this may be confined to acid resistant footwear and gloves. In other operations, where there is a greater possibility of acid leakage, full protective clothing must be worn.
  - A plentiful supply of water must always be available. A bath or shower adjacent to the work area is an advantage. Tepid water should be supplied so that any possibility of shock is reduced to a minimum. The shower must be equipped with a foolproof and easily operated valve or switching-on device.
  - Eyewash facilities must always be available and located in prominent and readily accessible positions.
  - Where acids may come into contact with metal, smoking or naked lights must be prohibited because of the possibility of hydrogen evolution.
  - For the same reason, any mild steel drums containing acid should always be regularly vented to prevent the build-up of pressure.
  - Applying pressure to the container should never discharge acid from carboys. Discharge should be by pouring or siphoning.

10.2 Electrical safety

This system requires a supply of AC electrical power, which is utilized to perform the ammonium hydroxide process. Coming in contact with AC power may be injurious or result in death. Do not operate the system with equipment covers or panels removed. When maintenance or other operations require removal of these panels, turn off, lock out, and tag the main breaker.

10.3 First aid

These items have been discussed previously. They are repeated for emphasis and recommendation that training be given to all personnel involved with the system.

10.3.1 Eye Burns — acid and alkali materials

Irrigate the eye immediately, before moving the patient for medical treatment. If Neutralize or another approved eyewash is not available, use copious amounts of fresh water. Irrigate for at least 15 min, while holding the eyelids apart and rotating the eye. Seek immediate medical treatment.

10.3.2 Skin burns — sodium hypochlorite, acid, or alkali materials

Apply copious amounts of water to the burned area. Remove the affected clothing. Do not neutralize acid with an alkaline solution. Seek immediate medical assistance.

10.3.3 Ingestion or gassing — sodium hypochlorite or alkali materials

Obtain emergency medical assistance. If the patient has been gassed, remove him from the contaminated environment as soon as possible. If sodium hypochlorite or ammonium hydroxide has been swallowed, do not induce vomiting. Keep the patient warm and resting quietly. If breathing stops, and if trained, apply CPR and continue until released by a doctor.

10.3.4 Electrical shock

DO NOT TOUCH A VICTIM BEFORE SWITCHING OFF THE POWER SOURCE. If necessary and if trained, start CPR. Have someone call the emergency number, e.g., 911.

BRINE SYSTEM DATA SHEET

System serial number _____

Date _____

Name _____

Rectifier Data

Rectifier Clock Reading _____

AC Voltage   A _____ B _____ C _____

AC Current   A _____ B _____ C _____

DC Current meter   A _____ B _____ C _____

DC Voltage meter   A _____ B _____ C _____

Rectifier control Manual   Yes   No

                     Set point _____

                     Automatic setting _____

Feed Water Data

Inlet water temperature   _____

Water flow meter setting _____

Brine flow meter setting _____

Cell feed specific gravity _____

Product Data

Product temperature   _____

Product concentration   _____

SEAWATER SYSTEM DATA SHEET

System serial number _____

Date _____

Name _____

Rectifier Data

Rectifier Clock Reading _____

AC Voltage   A _____ B _____ C _____

AC Current   A _____ B _____ C _____

DC Current meter   A _____ B _____ C _____

DC Voltage meter   A _____ B _____ C _____

Rectifier control   Manual   Yes   No

                          Set point   _____

                          Automatic setting _____

Feed Water Data

Inlet water temperature _____

Flow meter setting   _____

Water specific gravity _____

Product Data

Product temperature   _____

Product concentration _____

Technical Standard Documents

*TS-01–TS-35**

* From Severn Trent Services, 2002. With permission.

Technical Standard Document No. TS-01

Standard Title: Transformer/Rectifier Specification

Revisions   Date   Prepared By   Approved By   Purpose

1.0 Design:
NOTE: The following specification is provided as a standard. Units being provided shall be in full compliance with this document unless an Exceltec International, hereafter referred to as EIC, client specification is provided. In this case, the client specification has precedence over areas which conflict. Supplier is to make EIC aware of all such conflicts and EIC will then instruct the supplier on how to proceed.

   1.1 Units shall be suitable for indoor use and shall be cooled by means of forced air, unless stated otherwise in the project specification list. These units are to be used in electrochemical duty, as defined in ANSI standard C34.2. The equipment defined in this specification will be installed in a Tropical Environment, and is expected to operate with minimum supervision and maintenance.

   1.2 Ratings:
Input: Per Project
Output: Per Project

   1.3 Rectification Circuit Design:
Vendor Standard

   1.4 Control:
      1.4.1 Constant current from 0 to 100% of rating shall be achieved by thyristor phase control using a 10 turn pot. Output shall be within (+/-) 1% over the combined effects of 50 to 100% DC output voltage at rated ambient temperature range and (+/-) 5% AC input line voltage variation.
      1.4.2 Triggering circuits shall be inherent timing balance within 50 microseconds, and shall be sensitive to power system distortion. Trigger shall inhibit output on the loss of an input phase.
      1.4.3 Gate trigger lockout shall be activated from either a front panel switch, or an interlocking contact from the process control system.

   1.5 Environment:
Continuous operation is to be expected under maximum ambient conditions of 40°C with 100% relative humidity. The units will be located indoors and may be shutdown for significant periods with high humidity. The above shall be considered standard unless stated otherwise in the project specification list.

Strip heaters shall be provided in each cabinet, which will be automatically energized on shutdown of the units. Protection against excessive heating shall be provided by means of a thermostat with a minimum range of 20°C to 60°C. Input power to the strip heaters shall be protected by circuit breakers. Space heaters shall utilize an independent power feed for supply and logic.

2.0 Construction:
  2.1 Transformer:
    2.1.1 Dry type air cooled, 130°C maximum rise.
    2.1.2 Insulation class H.
    2.1.3 Windings primary and secondary use electrolytic grade copper.
    2.1.4 Core — grain oriented steel, grade M6 or better.
    2.1.5 Coils — Vacuum impregnated, epoxy coated.
    2.1.6 All secondary connections shall be controlled in location, so that all like transformers are interchangeable, and a replacement may be installed at a later date with minimum interference.
    2.1.7 The transformer shall conform to the requirements of ANSI standards C57.12.01 and C57.18 for electrochemical duty.
    2.1.8 The transformer shall be permanently tagged with the transformer/rectifier unit serial number.
  2.2 Cabinet:
    2.2.1 Each cabinet shall be provided with a copper grounding lug of sufficient size.
    2.2.2 Transformer cubicle shall meet NEMA 1 standards, with sufficient removable hinged access doors to permit service of any component without prior disassembly of any other components. All ventilation openings shall be fitted with fine mesh screens to prevent the entrance of rodents, large insects, or debris. Top ventilation opening shall have means of connecting duct work for purpose of exhausting heated air.
    2.2.3 Control section shall meet NEMA 4X standards. The control space shall be of sufficient size as to facilitate service. Provision shall be made for the purchaser's cable/conduit entrance from either the top or bottom. Twenty percent of the panel space shall be left clear for future use. All components shall be attached in a manner in which they can be removed without removing the back pan.
    2.2.4 EIC paint specification TS-025 shall apply.
    2.2.5 Removable intake air filters that may be cleaned shall be provided
  2.3 Rectifier Section:
    2.3.1 Conductors shall be electrolytic grade copper of at least one square inch per 1250 Amps in each path for a forced air unit. For a sealed cooling unit, 1000 amps shall be utilized.
    2.3.2 Power semiconductors shall have blocking voltage ratings of a minimum of 1200 volts, but at least 10 times the maximum DC voltage output under any conditions. Suf-

ficient cooling shall be provided to keep the maximum case temperature at least 15°C below the maximum rating for the device at rated current and at rated maximum coolant temperature.

2.3.3 Secondary power semiconductors shall be individually fused with current limiting fuses of the appropriate rating to isolate a faulted path without further damage to the unit.

2.4 Cooling System:

2.4.1 Cooling of the power components is by forced air, unless otherwise stated. Power semiconductors shall be mounted on the properly sized, anodized or iridited aluminum heat sinks.

2.5 Wiring:

2.5.1 Control wiring shall be of industrial quality, rated at 105°C or better. Wire shall be rated for 600 V and the size shall be 14 awg or larger, except for thyristor gate leads. Conductor shall be fine stranded tinned copper and insulation shall be PVC. Wire shall be UL listed. Manhattan style #M216 is preferred.

2.5.2 Terminations shall be by full tongue compression type terminals that provide a gas-tight connection, and have insulated compression sleeves to grip the wire insulation, or industrial grade clamp type terminal strips. Alternatives may be utilized if approved by an EIC engineer.

2.5.3 All wiring shall be labeled with permanent wire number markers, with a consistent numbering scheme. All wire numbers shall be shown on the electrical drawings. Wire numbers for a series sequence of circuit elements shall increase by one for each element passed through.

2.5.4 No more than two wires shall be landed on any termination point.

2.5.5 Wire ducts shall be used to organize and route control wiring. Proper routing shall provide separation between AC control and signal wiring. Wire ducts shall be sized so that the enclosed wiring consumes no more than 50% of the total available space.

2.5.6 Power wiring shall be of a size and insulation level appropriate to its application.

2.5.7 Equipment should be assigned tag numbers and be indicated with engraved plastic tags (Gravopoly) and attached by stainless steel screws. Tags should be white with black lettering. Tag numbers should correspond with all drawings. Alternate tagging methods may be utilized if first approved by EIC.

2.6 Terminal Blocks:
 2.6.1 All wiring for external field connections shall be terminated on a minimum number of conveniently located terminal blocks.
 2.6.2 Terminal blocks shall be mounted so that sufficient clearance shall be provided for terminal marking, wire insertion and removal, and for the purchaser's wiring connection.
 2.6.3 Terminal blocks shall be of adequate size and shall be designed to receive purchaser's incoming control cables.
 2.6.4 A minimum of 20% spare positions shall be provided on all terminal blocks.
2.7 Controls:
 2.7.1 General service auxiliary control relays shall be of an octal base, plug in type sealed design. They shall be IDEC or equal. Only one contact of each form C pair is to be utilized. Vendor may utilize PLC logic if approved by EIC.
 2.7.2 Dry circuits (current < 50 mA, voltage < 30 V) must be switched by relays rated for the application, IDEC type RR3PA-UL or approved equal.
 2.7.3 Push buttons, switches and lights shall be industrial quality, oil tight construction suitable for NEMA 4X application.
 2.7.4 Electronic controls must be of printed circuit construction. They must also be mechanical locking, and have corrosion resistant contact materials.
 Any edge connecting circuit boards must have a system for mechanically securing the boards and have a corrosion resistant plating system for the connecting edge, such as Ni or Au. All electronic boards shall be completely tropicalized before installation. If any board adjustments are made during testing these areas should be touched up.
 2.7.5 Proper protection (i.e., fuse, circuit breakers) shall be provided for all equipment as dictated by the National Electric Code.
2.8 Tropicalization:
 All electrical equipment, internal bus bars, and their enclosures shall be tropicalized. Secondary wiring, coils, and other insulations that are not fungus resistant, shall have a fungus resistant coating applied, except where such coatings would interfere with the proper operation. In such cases, the part shall be inherently fungus resistant. Use Dolph's Synthite AC-279–7s Clear Air Drying Anti-fungal Varnish or EIC approved equal.
2.9 Motors:
 All motors shall be totally enclosed fan cooled (TEFC) rated for the design ambient conditions. Insulation shall be Class F Continuous Duty.

3.0 Protection and Metering:
  3.1 Design:
   Vendor shall be responsible for proper protection coordination and shall supply to EIC coordination curves.
  3.2 Protection:
   3.2.1 Circuit protection shall be supplied as follows:
    3.2.1.1 Fuses or an incoming power circuit breaker shall be a manually operated type with an under-voltage trip. Design shall be sized to correctly protect the transformer/rectifier.
    3.2.1.2 Auxiliary power shall be protected by fuses or circuit breakers which are manually operated type. Individual fuses or circuit breakers should be provided for the 110/120 VAC control power and for the space heaters.
    3.2.1.3 Circuit breakers shall be of the indicating type providing ON, OFF, and TRIPPED positions of the operating handle. Multiple breakers shall be designed so that an overload on one pole shall open all poles.
   3.2.2 Instantaneous AC over-currents in excess of 150% rated input current shall immediately inhibit thyristor firing pulses. Removal of the over-current condition shall allow a smooth ramp recovery of output.
  3.3 Alarms and Shutdowns:
   3.3.1 Each of the following conditions shall be indicated locally and cause shutdown. These indications shall be latching indications, which are reset by an acknowledge push-button. A shutdown condition causes the thyristor triggering to be inhibited, or the primary AC circuit breaker to be opened. A test push-button should also be provided for periodic testing of all lamp indicators. All alarm and shutdown circuits shall be designed using Fail Safe logic.
    3.3.1.1 AC over-current trip. A non-thermal, adjustable over-current relay shall be provided, with any required current transformers, which shall trip the unit for any input phase current greater than 15% above the continuous rating of the unit.
    3.3.1.2 Semiconductor fuse failure shutdown. Visual indication of individual failed devices shall be provided.
    3.3.1.3 Semiconductor over-temperature shutdown (one sensor for each semiconductor assembly).
    3.3.1.4 Transformer over-temperature shutdown (one sensor for each phase).
    3.3.1.5 DC over-voltage shutdown, set at 105% of rated output.

- 3.3.1.6 DC over-current shutdown, set at 105% of rated output.
- 3.3.1.7 Blower/Fan failure overload trip.
- 3.3.2 Remote indication shall be provided by means of a Form C isolated contact for each of the following functions.
  - 3.3.2.1 Rectifier On condition (AC Power Supplied).
  - 3.3.2.2 Tripped - Fault Relay Contacts.
- 3.4 Metering:

  The following meters shall be provided as a minimum on each unit. Meters shall have an accuracy of +/- 2% of full scale. All required instrument transformers shall be supplied.
  - 3.4.1 DC Ammeter.
  - 3.4.2 DC Voltmeter.
  - 3.4.3 Elapsed running time meter.
- 3.5 Panel Controls:

  The following panel controls will be mounted on the control cabinet of each individual unit:
  - 3.5.1 Auxiliary power (circuit breaker or fuses)
  - 3.5.2 Current control potentiometer. One turn unit with indicating knob, scaled 0–100% for 0 to rated output current.

4.0 Quality Control and Testing:
- 4.1 Documents:
  - 4.1.1 Project schedule. Within two weeks of receipt of order, the vendor shall provide an estimated project schedule showing the anticipated dates of major events in the project.
  - 4.1.2 Approval drawings. The following drawings are required for customer approval, before start of construction. For the purpose of scheduling, assume 8 weeks from submittal for drawing approval. One reproducible Mylar with a D size border and 1/8" text is required during each submittal stage.
    - 4.1.2.1 Mechanical outline drawing showing detailed cabinet outline with dimensions, power and control entry locations and dimensions, service access, control locations and panel layout, and provisions for moving the equipment.
    - 4.1.2.2 Electrical schematic drawings showing the close-up wiring, detailing all protective circuits and rectifier control logic.
    - 4.1.2.3 Bill of materials showing component specifications of all major components such as power semiconductors, protective relays, connectors, fuses, etc.
  - 4.1.3. Final drawings. One reproducible with D size border and at least 1/8" text shall be submitted of the following drawings.
    - 4.1.3.1 Mechanical outline and installation drawing.

4.1.3.2 Major component layout and service access drawing.
4.1.3.3 Internal schematic diagram(s) showing the power circuit, protection circuits, controls, customer interface, and metering circuits. All wire numbers shall be shown. Drawing formats and symbols shall comply with ISA Standards and IEEE Standards for electrical circuits.
4.1.3.4 Field installation wiring diagram, showing all required field terminations, terminal locations, terminal numbers, and wire sizes.
4.1.3.5 Electronic control schematic and layout drawing.
4.1.3.6 Instruction manuals. Specified copies shall be supplied, containing the following information. Note that EIC standard vendor requirement is for three (3) copies plus the amount specified by our customers. Past history has shown that 12–18 copies are required.
4.1.4 Installation, Operating, and Maintenance Manual. The specified number of copies shall be submitted after the draft copy is approved. Manuals should contain the following as a minimum.
4.1.4.1 Equipment description, layout and operation.
4.1.4.2 Maintenance instructions.
4.1.4.3 Electronic controls and testing.
4.1.4.4 Complete Bill of Materials showing manufacturers and ratings. Copies of manufacturer's catalog pages or data sheets must be supplied. Supplier's correct part numbers should be included.
4.1.4.5 Complete detailed test reports, as described elsewhere.
4.1.4.6 Certified As Built drawings as described in Section 4.1.3 should be supplied as part of this manual.
4.2 Inspection:
4.2.1 In-process inspection. EIC reserves the right to inspect progress on this project at any time.
4.2.2 Final inspection shall be witnessed by EIC, or their representative, at full load of one unit. The vendor shall have completed a full load heat run on all units prior to this witness test. Vendor shall notify EIC at least 2 weeks prior to that point of the test schedule. This inspection may be waived by EIC.
4.2.3 Test reports. Test procedures and report forms shall be submitted in detail to EIC for approval prior to testing transformers and rectifiers. At least 2 weeks shall be allowed for this approval process.

4.2.4 Final test reports shall be submitted to EIC after inspection and test.
4.3 Transformer tests:
   4.3.1 Transformer tests shall be done in accordance with ANSI standards C57.12.01 and C57.18. Test reports on each transformer shall be traceable to the unit in which it is installed. A copy of the transformer test report shall be included in the instruction manual for that unit.
   4.3.2 Type tests. The following tests shall be performed and noted on the test report on one transformer, prior to installation in the rectifier units.
      4.3.2.1 Excitation loss.
      4.3.2.2 Copper loss.
   4.3.3 Unit tests. Each transformer shall be given the following tests prior to installation in the rectifier units. These tests are to be noted on the test report and shall be traceable to the transformer/rectifier serial number.
      4.3.3.1 Dielectric test.
      4.3.3.2 Ratio test.
      4.3.3.3 Noise level should not exceed 85 dBA. Supplier should guarantee this in writing.
4.4 Electronics:
   4.4.1 Burn in tests. Each electronic control board shall be given an operating, 2-hour burn in test, at an ambient temperature of 70°C. Alternate burn in procedures may be acceptable if submitted for approval.
   4.4.2 Documented tests conforming to written procedure approved by EIC shall be performed on each circuit board, both before and after the burn in test. Some method of board traceability must be provided.
   4.4.3 Traceable calibration data shall be provided for any control requiring calibration.
4.5 Unit Tests:
   4.5.1 The following tests shall be conducted in accordance with ANSI standard C34.2. Copies of all test reports shall be included in the instruction manual.
   4.5.2 Type tests: These tests shall be conducted on one unit of each design, prior to the witness test and shall be documented on the general test report.
      4.5.2.1 Power losses.
   4.5.3 Unit tests: These tests shall be conducted on each unit, and documented on the unit test report.
      4.5.3.1 Dielectric tests on each independent circuit section, i.e., high voltage, DC output, AC control, protection circuits, etc.
      4.5.3.2 Functional tests on all protective and control circuits.

4.5.3.3 Full load test — Semiconductor case temperatures are probed and recorded after one hour of operation.

4.5.3.4 Full current and full voltage tests per ANSI standard C34.2.

4.5.3.5 Temperature rise at full load. Temperature recordings shall be taken of transformer windings, semiconductor cases, magnetic cores, and the control environment.

Technical Standard Document No. TS-01.1

Standard Title: Tap Transformer/Rectifier Specification

Revisions   Date   Prepared By   Approved By   Purpose

1.0 Design:
NOTE: The following specification is provided as a standard. Units being provided shall be in full compliance with this document unless an Exceltec International, hereafter referred to as EIC, client specification is provided. In this case, the client specification has precedence over areas which conflict. Supplier is to make EIC aware of all such conflicts and EIC will then instruct the supplier on how to proceed.

1.1 Units shall be suitable for indoor use and shall be cooled by means of forced air, unless stated otherwise in the project specification list. These units are to be used in electrochemical duty, as defined in ANSI standard C34.2. The equipment defined in this specification will be installed in a Tropical Environment, and is expected to operate with minimum supervision and maintenance.

1.2 Ratings:
Input: Per Project
Output: Per Project

1.3 Control:
Current control from 0 to 100% of rating shall be achieved by AC primary transformer tap switches. Output DC current shall be within (+/-) 5% of setting at system operating equilibrium. DC voltage shall float according to system requirements.

1.4 Environment:
Continuous operation is to be expected under maximum ambient conditions of 40°C with 100% relative humidity. The units will be located indoors and may be shut down for significant periods of high humidity. The above shall be considered standard unless stated otherwise in the project specification list.

2.0 Construction:
2.1 Transformer:
2.1.1 Dry type air cooled, 130°C maximum rise.
2.1.2 Insulation class H.
2.1.3 Windings, primary and secondary, shall use electrolytic grade copper.
2.1.4 Core — grain oriented steel, grade M6 or better.
2.1.5 Coils — Vacuum impregnated, epoxy coated.
2.1.6 All secondary connections shall be controlled in location, so that all like transformers are interchangeable, and a replacement may be installed at a later date with minimum interference.
2.1.7 The transformer shall conform to the requirements of ANSI standards C57.12.01 and C57.18 for electrochemical duty.

2.2 Cabinet:
2.2.1 Each cabinet shall be provided with a copper grounding lug of sufficient size.

2.2.2 Transformer cubicle shall meet NEMA 1 standards, with sufficient removable hinged access doors to permit service of any component without prior disassembly of any other components. All ventilation openings shall be fitted with fine mesh screens to prevent the entrance of rodents, large insects, or debris. Top ventilation opening shall have means of connecting duct work for purpose of exhausting heated air.

2.2.3 Control section shall meet NEMA 12 standards. The control space shall be of sufficient size as to facilitate service. Provision shall be made for the purchaser's cable/conduit entrance from either the top or bottom. Twenty percent of the panel space shall be left clear for future use. All components shall be attached in a manner in which they can be removed without removing the backpan.

2.2.4 EIC paint specification shall apply:
Sandblast:
SSPC 10–85, Near White,
Less than 1 1/2 mil profile.
Primer Coat:
Red Lead Alkyd Primer, 2–3 mils DFT
(Carboline GP-30 or equal)
Intermediate Coat:
Red Lead Alkyd Primer, 1 1/2–2 mils
DFT tinted with carbon black or
lampblack to a color contrasting with
that of the primer.
(Carboline GP-30 or equal)
Finish Coat:
Modified Long Oil Alkyd, 1–2 mils
DFT (Carboline AD-51 or equal)
(Color Gray - C703)

2.3 Rectifier Section:

2.3.1 Conductors shall be electrolytic grade copper of at least one square inch per 1250 Amps in each path for a forced air unit. For a sealed cooling unit, 1000 amps shall be utilized.

2.3.2 Power semiconductors shall have blocking voltage ratings of a minimum of 200 volts, but at least 10 times the maximum DC voltage output under any conditions. Sufficient cooling shall be provided to keep the maximum case temperature at no higher than 15°C below the maximum rating for the device at rated current and at rated maximum coolant temperature.

2.3.3 Secondary power semiconductors shall be individually fused with current limiting fuses of the appropriate rating

to isolate a faulted path without further damage to the unit.
2.4 Cooling System:
   2.4.1 Cooling of the power components is by forced air, unless otherwise stated. Power semiconductors shall be mounted on the properly sized, anodized or iridited aluminum heat sinks.
2.5 Wiring:
   2.5.1 Control wiring shall be of industrial quality, rated at 105°C or better. Wire shall be rated for 600 V and the size shall be 14 awg or better. Thyristor gate leads. Conductor shall be fine stranded tinned copper and insulation shall be PVC. Wire shall be UL listed.
   2.5.2 Terminations shall be by full tongue compression type terminals that provide a gas tight connection, and have insulated compression sleeves to grip the wire insulation, or industrial grade clamp type terminal strips. Alternatives may be utilized if approved by an EIC engineer.
   2.5.3 All wiring shall be labeled with permanent wire number markers, with a consistent numbering scheme. All wire numbers shall be shown on the electrical drawings. Wire numbers for a series sequence of circuit elements shall increase by one for each element passed through.
   2.5.4 No more than two wires shall be landed on any termination point.
   2.5.5 Wire ducts shall be used to organize and route control wiring. Proper routing shall provide separation between AC control and signal wiring. Wire ducts shall be sized so that the enclosed wiring consumes no more than 70% of the total available space.
   2.5.6 Power wiring shall be of a size and insulation level appropriate to its application.
   2.5.7 Equipment should be assigned tag numbers and be indicated with engraved plastic tags (Gravopoly). Tags should be white with black lettering. Tag numbers should correspond with all drawings. Alternate tagging methods may be utilized if first approved by an EIC engineer. Tag fastening shall be by s.s. screws.
2.6 Terminal Blocks:
   2.6.1 All wiring for external field connections shall be terminated on a minimum number of conveniently located terminal blocks.
   2.6.2 Terminal blocks shall be mounted so that sufficient clearance shall be provided for terminal marking, insertion and removal, and for the purchaser's wiring connection.

2.6.3 Terminal blocks shall be of adequate size and shall be designed to receive purchaser's incoming control cables.
2.6.4 A minimum of 20% spares shall be provided on all terminal blocks.
2.7 Controls:
2.7.1 General Service auxiliary control relays shall be of an octal base, plug in type sealed design. They shall be IDEC or equal. Only one contact of each form C pair is to be utilized. Vendor may utilize PLC logic if approved by an EIC engineer.
2.7.2 Dry circuits (current < 50 mA, voltage < 30 V) must be switched by relays rated for the application, IEDC type RR3PA-UL or approved equal.
2.7.3 Push buttons, switches and lights shall be industrial quality, oil tight construction suitable for NEMA 4 application.
2.7.4 Proper protection (i.e., fuse, circuit breakers) shall be provided for all equipment as dictated by the National Electric Code.
2.8 Tropicalization:
All electrical equipment, internal bus bars and their enclosures shall be tropicalized. Secondary wiring, coils and other insulations that are not fungus resistant, shall have a fungus resistant coating applied, except where such coatings would interfere with the proper operation. In such cases, the part shall be inherently fungus resistant. Use Dolph's Synthite AC-279–7s Clear Air Drying Anti-fungal Varnish or EIC engineering approved equivalent.
2.9 Motors:
All motors shall be totally enclosed fan cooled (TEFC) rated for the design ambient conditions. Insulation shall be Class F Continuous Duty.
3.0 Protection and Metering:
3.1 Design:
Vendor shall be responsible for proper protection coordination and shall supply to EIC coordination curves.
3.2 Protection:
3.2.1 Circuit breakers shall be supplied as follows:
3.2.1.1 Incoming power circuit breaker shall be a manually operated type with an under-voltage trip. Circuit breaker should be sized to correctly protect the transformer/rectifier.
3.2.1.2 Auxiliary power circuit breakers shall be a manually operated type. Individual breakers should be provided for the 110 control power and for the space heaters.
3.2.1.3 Circuit breakers shall be of the indicating type providing ON, OFF, and TRIPPED positions of the

operating handle. Multiple breakers shall be designed so that an overload on one pole shall open all poles.
- 3.3 Alarms and Shutdowns:
  - 3.3.1 Each of the following conditions shall be indicated locally and cause shutdown. These indications shall be latching indications, which are reset by an acknowledge push-button. A shutdown condition causes the primary AC circuit breaker to be opened. A test push-button should also be provided for periodic testing of all lighted indicators. All alarm and shutdown circuits shall be designed using Fail Safe logic.
    - 3.3.1.1 Semiconductor over-temperature shut-down (one sensor for each semiconductor assembly).
    - 3.3.1.2 Transformer over-temperature shut-down (one sensor for each phase).
    - 3.3.1.3 DC Over-voltage shutdown, set at 105% of rated output.
    - 3.3.1.4 DC Over-current shutdown, set at 105% of rated output.
    - 3.3.1.5 Blower/Fan failure overload trip.
  - 3.3.2 Remote indication shall be provided by means of a Form C isolated contact for each of the following functions.
    - 3.3.2.1 Rectifier On condition (breaker closed).
    - 3.3.2.2 Tripped - Fault Relay Contacts.
- 3.4 Metering:

The following meters shall be provided as a minimum on each unit. Meters shall have an accuracy of +/- 2% of full scale.
  - 3.4.1 DC Ammeter.
  - 3.4.2 DC Voltmeter.
  - 3.4.3 Elapsed running time meter.
- 3.5 Panel Controls:

The following panel controls will be mounted on the control cabinet of each individual unit.
  - 3.5.1 Auxiliary power switch (circuit breaker).
  - 3.5.2 A.C. line 8 position tap switches for current control.

4.0 Quality Control and Testing:
- 4.1 Documents:
  - 4.1.1 Project schedule. Within 2 weeks of receipt of order, the vendor shall provide an estimated project schedule showing the anticipated dates of major events in the project.
  - 4.1.2 Approval drawings. The following drawings are required for customer approval, before start of construction. For the purpose of scheduling, assume 8 weeks from submittal for drawing approval. One reproducible Mylar with a D

size border and 1/8" text is required during each submittal stage.
- 4.1.2.1 Mechanical outline drawing showing detailed cabinet outline with dimensions, power and control entry locations and dimensions, service access, control locations and panel layout, and provisions for moving the equipment.
- 4.1.2.2 Electrical schematic drawings showing the overall wiring. detailing all protective circuits and rectifier control logic.
- 4.1.2.3 Bill of materials showing component specifications of all major components such as power semiconductors, protective relays, connectors, fuses, etc.

4.1.3 Final drawings. One reproducible with D size border and at least 1/8" text shall be submitted of the following drawings.
- 4.1.3.1 Mechanical outline and installation drawing.
- 4.1.3.2 Major component layout and service access drawing.
- 4.1.3.3 Internal schematic diagram(s) showing the power circuit, protection circuits, controls, customer interface, and metering circuits. All wire numbers shall be shown. Drawing formats and symbols shall comply with industry accepted standards.
- 4.1.3.4 Field installation wiring diagram, showing all required field terminations, terminal locations, terminal numbers, and wire sizes.
- 4.1.3.5 Electronic control schematic and layout drawing.
- 4.1.3.6 Instruction manuals. Specified copies shall be supplied, containing the following information. Note that EIC standard vendor requirement is for three (3) copies plus the amount specified by our customers. Past history has shown that 12–18 copies are required.

4.1.4 Installation, Operating, and Maintenance Manual. Specified copies shall be submitted after the draft copy is approved. Manuals should contain the following as a minimum.
- 4.1.4.1 Equipment description, layout and operation.
- 4.1.4.2 Maintenance instructions.
- 4.1.4.3 Electronic controls and testing.
- 4.1.4.4 Complete Bill of Materials showing manufacturers and ratings. Copies of manufacturer's catalog pages should be supplied where available. Supplier's correct part numbers should be included where applicable.

4.1.4.5 Complete detailed test reports, as described elsewhere.

4.1.4.6 Certified As Built drawings as described in Section 4.1.3 should be supplied as part of this manual.

4.2 Inspection:

4.2.1 In process inspection. EIC reserves the right to inspect progress on this project at any time.

4.2.2 Final inspection shall be witnessed by EIC, or their representative, at full load of one unit. The vendor shall have completed a full load heat run on said units prior to this witness test. Vendor shall notify EIC at least 2 weeks prior to that point of the test schedule. This inspection may be waived by EIC if desired.

4.2.3 Test reports. Test procedures and report forms shall be submitted in detail to EIC for approval prior to testing transformers and rectifiers. At least 2 weeks shall be allowed for this approval process.

4.2.4 Final test reports shall be submitted to EIC after inspection and test.

4.3 Transformer tests:

4.3.1 Transformer tests shall be done in accordance with ANSI standards C57.12.01 and C57.18. Test reports on each transformer shall be traceable to the unit in which it is installed. A copy of the transformer test report shall be included in the instruction manual for that unit.

4.3.2 Type tests. The following tests shall be performed on one of the transformers, prior to installation in the rectifier units.

4.3.2.1 Excitation loss.

4.3.2.2 Copper loss.

4.3.3 Unit tests. Each transformer shall be given the following tests prior to installation in the rectifier units.

4.3.3.1 Dielectric test.

4.3.3.2 Ratio test.

4.3.3.3 Noise level should not exceed 85 dBA. Supplier shall guarantee this in writing.

4.4 Unit Tests:

4.4.1 The following tests shall be conducted in accordance with ANSI standard C34.2. Copies of all test reports shall be included in the instruction manual.

4.4.2 Type tests: These tests shall be conducted on one unit of each design, prior to the witness test.

4.4.2.1 Power losses.

4.4.2.2 Temperature rise at full load. Temperature recordings shall be taken of transformer windings, semi-

conductor cases, magnetic cores, and the control environment.
4.4.3 Unit tests: These tests shall be conducted on each unit, and documented on the unit test report.
- 4.4.3.1 Dielectric tests on each independent circuit section, i.e., high voltage, DC output, AC control, protection circuits, etc.
- 4.4.3.2 Functional tests on all protective and control circuits.
- 4.4.3.3 Full load test — one hour reading is to be recorded, semiconductor case temperatures are to be probed and recorded.
- 4.4.3.4 Full current and full voltage tests per ANSI standard C34.2.

Technical Standard Document No. TS-03

Standard Title: Electrolyzer Module Specification

| Revisions | Date | Prepared By | Approved By | Purpose |

1.0 Design:
NOTE: The following specification is provided as a standard. Units being provided shall be in full compliance with this document unless an Exceltec International, hereafter referred to as EIC, client specification is provided. In this case, the client specification has precedence over areas which conflict. Supplier is to make EIC aware of all such conflicts and EIC will then instruct the supplier on how to proceed.

2.0 Electrolyzer Module:
   2.1 This specification covers the electrolytic cells, interconnecting piping and valves, bus-work, and all other equipment that is necessary to make the production streams complete, contributing to smooth reliable plant operation.
   2.2 The design of this plant ensures reliable operation in a chemical and corrosive environment. The module frames shall be constructed of flame retardant extruded fiberglass or epoxy coated steel structural shapes.
   2.3 The electrolytic cell module assemblies shall be supplied with seawater from a common header. The seawater shall be pressurized to a minimum of three (3) bar g and pumped through strainer(s) to remove particles larger than 0.8 mm. The sodium hypochlorite solution produced shall be sent to the injection points after removal of entrained hydrogen gas.
   2.4 The flow of seawater through the electrolytic modules shall be kept at a constant rate and sufficient to minimize the build-up of precipitated calcium and magnesium salts on the electrodes.
   2.5 All cells shall be equipped with acrylic covers to allow easy monitoring of deposit build-up that occurs on the cathodes due to the calcium and magnesium in the seawater.
   2.6 All cell assemblies shall be a monopolar plate type designed and constructed to enable easy removal and replacement of cell electrodes with due regard to the need for inspection and maintenance.
   2.7 The anodes and cathodes shall be spaced uniformly with a gap of 2.5 mm providing optimum cell efficiency with a low water pressure drop. Internal spacers shall be placed strategically over the electrode surface to maintain a uniform gap. Spacer material shall be Kynar® (polyvinyl difluoride) to withstand the turbulent corrosive environment in which they operate.
   2.8 Cell anodes shall be expanded titanium metal with a precious metal oxide coating and operated at a current density of less than 1600 amps/m^2. They shall be dimensionally stable type and shall preserve their shape and voltage characteristics under severe electrolytic conditions.
   2.9 Cell assembly cathodes shall be constructed of a Hastelloy C-276 material to offer superior corrosion resistance to seawater

and sodium hypochlorite while improving power consumption by 10% comparative to the use of titanium cathodes.

2.10 The cell housing shall be injection-molded polypropylene with ultraviolet stabilizers and integrally molded flanges to ensure positive sealing.

2.11 The maximum applied voltage differential (per cell) shall not exceed 7 V DC during the operation. A seawater temperature of 25 °C with a chloride content of 18,980 ppm should be used for this basis.

2.12 All internal cell assembly fasteners shall be commercially pure titanium. All external cell assembly fasteners shall be brass or silicone bronze. All cell assembly gaskets or seals shall be either Viton or silicone rubber.

2.13 Interconnecting cell assembly bus bar shall be constructed from 99.8% pure electrolytic grade copper.

2.14 All piping shall be Schedule 80 PVC with ultraviolet stabilizers, Type 1, Grade 1. All PVC Schedule 80 pipe shall conform to ASTM D-1784 and D-1785 standards. Schedule 80 socket fittings shall meet ASTM D-2467 while threaded fittings shall conform to ASTM D-2464. Both pipe and fittings shall be the product of one manufacturer.

2.15 Valves shall be provided as follows:

2.16 Ball Valves: Shall be Schedule 80 PVC, True Union type with Viton O-Ring seals and Teflon seats.

2.17 Butterfly Valves: Shall be Schedule 80 PVC with 316SS shafts, EPDM seats and shaft seals. Fasteners shall be 316 stainless steel.

2.18 Actuators: Ball or butterfly valves shall be equipped with pneumatic operated or electric actuators. Permanent design shall be Fail-Spring to Close. Actuator housings should be NEMA 4 construction and should contain two (2) limit switches to be used for remote indication.

3.0 Electrolyzer Module Instruments:

The electrolyzer module shall contain as a minimum the following instruments:

1. Inlet Low Flow Indicating Switch:

Type:	Rotary; Electro-mechanical basis.
Accuracy:	+/- 2% of full range
Repeatability:	+/- 0.5% of full range
Construction:	PVDF with Hastelloy C shaft
Indication:	Analog with Low Flow Switch
Switch Contact:	SPDT, 3 A at 240 V AC
Enclosure:	FRP Panel with Clear Cover

2. Pressure Relief Valve:

Type:	Schedule 80 PVC; Type 1, Grade 1; Angle Pattern Design
Seals:	Viton
Shaft:	Teflon
Fasteners:	316 Stainless Steel
Spring:	Stainless Steel

3. Pressure Gauge (Inlet and Outlet):

Type:	Process Gauge — Liquid Filled
Case Material:	Glass Reinforced Thermoplastic or S.S.
Window Material:	Acrylic
Dial Size:	4.5"
Accuracy:	0.5% of Full Range
Bourdon Tube:	316 Stainless Steel
Scale:	Dual Scale — PSI and Bar
Connection:	Lower Mount —1/4" NPT
Safety Feature:	Solid Front — Blow Out Back
Gauge Guard:	PVC Type — 1/2" NPT Process connection 1/4" NPT Instrument connection

4. Outlet Low Flow Switch:

Type:	Paddle Actuator
Construction:	Ryton body; Viton seats; Hastelloy C wetted parts
Switch Contact:	SPDT, 15 A at 240 V A.C.

5. Temperature Gauge (outlet): (NOTE: As required)

Type:	Bi-metal Thermometer — Liquid Filled
Case Material:	Glass Reinforced Thermoplastic
Window Material:	Acrylic
Dial Size:	5"
Accuracy:	1% of Full Span
Scale:	deg C
Connection:	Every angle — 1/2" NPT

Technical Standard Document No. TS-04

Standard Title: Centrifugal Pump Specification

Revisions	Date	Prepared By	Approved By	Purpose

1.0 General:

NOTE: The following specification is provided as a standard. Proposals being submitted shall be in full compliance with this document unless an Exceltec International, hereafter referred to as EIC, client specification is provided. In this case, the client specification has precedence over areas which conflict. Supplier is to make EIC aware of all such conflicts and EIC will then instruct the supplier on how to proceed.

2.0 Materials of Construction:
- 2.1 The cover, casing, gland, base plate, and impellor shall be constructed of fiberglass-reinforced Derakane vinyl ester resin, utilizing continuous strand fiberglass mat.
- 2.2 The shaft and fasteners shall be constructed of 303 SS or equal.
- 2.3 The mechanical seal shall be constructed with carbon and chemical grade ceramic faces, 316 SS parts and Viton secondary seals, seal code E75 — V or equal.
- 2.4 Gaskets are to be constructed of Viton or equal.
- 2.5 The adapter, bearing frame, and bearing housing shall be constructed of cast iron with an epoxy coating (Carboline 198 recommended).

3.0 Mechanical Construction:
- 3.1 Pumps shall be end-suction, top discharge type constructed IAW ANSI B73.1 dimensional standards. They shall have back pull out construction to facilitate maintenance, mounted on a common base with the motor, complete with woods type couplings and OSHA approved coupling guards.
- 3.2 The casing shall be one-piece construction with integral 150# ANSI flanges. Casings on 10" (nominal impeller size) and larger pumps shall be dual volute to reduce radial thrust. The impeller shall be semi-open with both pump out vanes and balance holes to achieve axial balance. The impeller shall have a low stress Hastelloy insert to provide full metal-to-metal contact with the shaft. The shaft seal shall be integral with the impeller to prevent damaging leakage past shaft sleeve o-rings.
- 3.3 Durametallic RAC or equal shaft seals shall be provided with an internal flush path to eliminate the need for an external source of flushing fluid.
- 3.4 The shaft shall be designed to ensure that deflection at the stuffing box is limited to .002" or less when the pump is operating at shutoff head to maximize seal and bearing life. The bearing housing shall be arranged to provide externally accessible impellor adjustment.
- 3.5 The base plate shall be ANSI dimensioned with a sloped center collection drain under the pump with a tapped drain hole. Base plates shall have a grout fill hole provided. Wall thickness shall be suitable to permit drilling for pump and motor mounting

bolts. Corrosion vulnerable metal inserts or stiffening plates are not permitted.

4.0 Motors:

Motors shall be squirrel-cage induction type suitable for operation in hostile environments. Frame, bearing housing, fan housing, and conduit box shall be constructed of cast iron and coated with corrosion resistant epoxy enamel paint. Motors shall be NEMA B design with ratings based on class B temperature rise above an ambient of 40°C with class F insulation. Service factor shall be 1.15 and sized to be non-overloading over the entire pump curve. All motor bearings shall be anti-friction type and be re-greaseable.

5.0 Documentation:

    5.1 The following documents shall be available as a minimum by the supplier:

        1) Dimensioned outline drawing
        2) Sectional drawing
        3) Bill of materials
        4) Mechanical seal outline drawing
        5) Performance curves
        6) Installation, operating, and maintenance manual
        7) Factory standard test reports

Technical Standard Document No. TS-05

Standard Title: Self-Cleaning Strainer Specification

Revisions   Date   Prepared By   Approved By   Purpose

1.0 General:

NOTE: The following specification is provided as a standard. Proposals being submitted shall be in full compliance with this document unless an Exceltec International, hereafter referred to as EIC, client specification is provided. In this case, the client specification has precedence over areas which conflict. Supplier is to make EIC aware of all such conflicts and EIC will then instruct the supplier on how to proceed.

Strainer shall be automatic self-cleaning type and shall be fabricated to ANSI B34.1 standards. It shall be designed to operate at preset timed intervals and also include a high differential pressure switch which will override the timed sequence.

2.0 Materials of Construction:

The body and cover shall be fabricated from 316 SS or equal and the straining element shall be constructed of Monel or equal.

3.0 Mechanical Characteristics:

The strainer shall be sized to handle 110% of the required flow rate with a pressure loss of less than 2 psig (.3 bar) with clean straining elements. The straining elements shall be of the reverse rolled slotted wedge wire design with 1/32" (800 micron) openings. The wide or flat cross section of the wedge wire screen shall face the direction of flow providing a smooth flat surface to trap debris. The reverse rolled slotted wedge shall be free of pockets, tubes, collector bars, etc. that accumulate and trap debris permanently. The strainer element must be able to withstand high differential pressure without failure or distortion.

The strainer shall be provided with drive shaft and hollow port assembly fitted with all necessary bearings and seals. The drive shaft and hollow port assembly will be free running at a maximum speed of two (2) revolutions per minute and shall not contact the screen surface. The port assembly shall be factory and field adjustable for positive, effective cleaning and shear capability. The drive shall be supported at the top with roller bearings located in a double reduction gear reducer. The bottom shall be supported with a water lubricated guide bearing.

The gear reducer shall be driven by a TEFC motor at the specified three phase voltage.

4.0 Strainer Controls:

A backwash control package shall be provided for continuous or intermittent operation of the backwash cycle. All controls shall be located in the main system control panel or a local NEMA 4X panel. Control package shall contain all necessary components for the full and proper operation of the strainer. Indication lights shall also be provided for Power On, High Differential Pressure, and Backwash Operating as a minimum. Industrial type components shall be used. A single element differential pressure indicating switch shall be used to override the timed backwash cycle and an alarm.

Technical Standard Document No. TS-05A

Standard Title: Manual Strainer Specification

Revisions   Date   Prepared By   Approved By   Purpose

1.0 Applicability:
NOTE: The following specification is provided as a standard. Proposals being submitted shall be in full compliance with this document unless an Exceltec International, hereafter referred to as EIC, client specification is provided. In this case, the client specification has precedence over areas which conflict. Supplier is to make EIC aware of all such conflicts and EIC will then instruct the supplier on how to proceed.

This specification applies to simplex basket type strainers installed on the inlet to the electrolyzer.

2.0. Materials of Construction:
The strainer body shall be constructed of 316 SS or equal and the basket shall be constructed of 316 SS, monel, or equal.

3.0 Mechanical Characteristics:
The strainer shall be constructed to ANSI B31.4 standards and shall be fitted with 150# ANSI flanges for installation in the system. It shall be designed for 150 psig (9.375 bar) at 150°F (65°C). The strainer shall be fitted with vent and drain connections and a blind flange bolted cover. The strainer shall be sized to handle 110% of the required flow rate with a pressure loss of less than 2 psig with a clean basket. The straining element shall be designed with 1/32" (800 micron) openings in a wedge wire element design.

Technical Standard Document No. TS-06

Standard Title: Logic Diagram Specification

Revisions   Date   Prepared By   Approved By   Purpose

1.0 Design:

NOTE: The following specification is provided as a standard. Drawings being submitted shall be in full compliance with this document unless an Exceltec International, hereafter referred to as EIC, client specification is provided. In this case, the client specification has precedence over areas which conflict. Supplier is to make EIC aware of all such conflicts and EIC will then instruct the supplier on how to proceed.

1.1 State:

Logic systems shall be of the energized design type where the main fault relay is de-energized to shut down the system. This is accomplished by having the fault switch open when the shutdown condition is reached, thereby de-energizing a relay and opening a contact in series with the fault relay. This is characterized by having the fault contacts in series.

2.0 Drawings:

2.1 Symbols:

Drawings shall be made using the standard symbols and abbreviations found in ANSI/IEEE STD 91–1984. If there are any deviations or if the possibility of confusion exists, the supplier shall provide a legend with the logic diagram.

2.2 System Condition:

The system shall be shown in the normal shutdown condition with no power applied. All switches, relays, contacts, and other components shall be in their expected conditions for this operating condition.

2.3 Labeling:

All lines in the ladder diagram shall have reference numbers applied to the left margin. Each line, where applicable, shall have the circuit name applied to the right margin, i.e., Seawater Strainer High Differential Pressure along with a line number location of the contacts controlled by the relay, with the normally closed contacts underlined.

All switches shall be labeled with a noun name and switch identification number and the set point should also be included where applicable. Switch position shall also be included on the drawing.

All other components shall be labeled with an identification number and, if practical, a noun name and component rating.

2.4 Submittal:

All logic diagrams shall be submitted on reproducible 22 x 34 Velum (or equal) and shall include a complete bill of materials and a legend (if applicable).

Technical Standard Document No. TS-07

Standard Title: Storage Tank Specification

Revisions   Date   Prepared By   Approved By   Purpose

1.0 General:
NOTE: The following specification is provided as a standard. Proposals being submitted shall be in full compliance with this document unless an Exceltec International, hereafter referred to as EIC, client specification is provided. In this case, the client specification has precedence over areas which conflict. Supplier is to make EIC aware of all such conflicts and EIC will then instruct the supplier on how to proceed.
All FRP tanks shall conform to the applicable sections of ASTM D 3299 - 81.

2.0 Materials of Construction:
The primary material of construction for all acid storage or neutralization tanks shall be high density polyethylene. Hypochlorite storage tanks shall be FRP using Dow 411 vinyl ester resin. Tanks for other purposes shall be considered as hypochlorite storage tanks for the purpose of this standard.

3.0 Design:
   3.1 Flanges:
   All connections to tanks shall be made via ANSI 150# flanges.
   3.2 Shape:
   All tanks shall be cylindrical with flat bottoms, dome tops, and mounted vertically.
   3.3 Design Conditions:
   Acid storage and neutralization tanks shall be designed for 0 psig at 140°F. Hypochlorite tanks shall be designed for 0 psig at 150°F. All polyethylene tanks shall be designed to ASTM D-1784 - 75 standards and FRP tanks shall be designed to ASTM D-3299 - 81 standards.
   3.4 Fittings:
   All tanks over 4' diameter shall have a side man-way installed. Smaller tanks shall have a means of cleaning provided, either through a hand-hole, removable cover, or any other standard practice. Tanks shall also have lifting means and anchoring provisions suitable to withstand 100 mi/h wind velocities.

4.0 Documentation:
Vendor shall provide sufficient information with each tank for EIC to produce standard detail drawings for the tank.

Technical Standard Document No. TS-08

Standard Title: Control Panel Specification

Revisions   Date   Prepared By   Approved By   Purpose

1.0 Applicability:
NOTE: The following specification is provided as a standard. Proposals being submitted shall be in full compliance with this document unless an Exceltec International, hereafter referred to as EIC, client specification is provided. In this case, the client specification has precedence over areas which conflict. Supplier is to make EIC aware of all such conflicts and EIC will then instruct the supplier on how to proceed.
This specification is applicable to freestanding control panels not built into the transformer/rectifier assembly.

2.0 Materials of Construction:
   2.1 Cabinet:
The control cabinet shall be constructed of 12-gauge steel or thicker.

3.0 Design:
   3.1 Style:
The control panel assembly shall be a freestanding frame type panel.
   3.2 Construction:
The panel shall be, as a minimum, NEMA type 4/IEC IP-55 outdoor cabinet with full fillet welded edges on all exterior surfaces. Where two or more front sections are required, the interior shall be an open frame construction without separating partitions. Removable lifting rings shall be installed in the top gusset plates.
   3.3 Panel finish:
Unless otherwise specified, the panel shall be finished in accordance with EIC standard paint specifications.
   3.4 Nameplates:
The location of all instruments, pushbuttons, etc., mounted in the front of the panel shall be identified by laminated plastic nameplates attached with stainless steel screws. Nameplates should be, where possible, mounted below the instrument. Internal instruments should have a similar nameplate which may be attached with two part epoxy.
   3.5 Process Piping:
No process piping of any kind is allowed in the panel. The only type of piping allowed is for instrument air.

4.0 Instrument Air System:
   4.1 Instrument Air System Piping:
The pneumatic piping shall be installed such that all instruments are accessible for adjustment, calibration, and removal. All threaded connections in the pneumatic system shall be ANSI B 1.20.1 tapered.
   4.2 Air Supply:
The air header shall be sized for the capacity of all instruments and control functions with an extra 20% capacity for possible

future installations. The header shall be constructed of 316 SS with Stainless Steel take off nipples mounted on the top of the header for connection of the instruments. Each instrument shall be equipped with 1/4 FPT isolation valves. A header pressure relief valve with a lifting lever shall be installed to prevent overpressure of the instruments.

4.3 Air Filtration:
An air reducing station with air filter, pressure regulator, and pressure gauge is required.

4.4 Tubing:
Tubing shall be 1/4" 316 SS or equal. Fittings shall be 316 SS Swagelok or equal. All incoming or outgoing pneumatic signal lines shall terminate in bulkhead fittings.

5.0 Wiring:

5.1 Wire Installation:
Electrical wiring inside of the panel shall conform to the requirements of the National Electrical Code, NFPA 70.

5.2 Wire Size:
General wiring shall be at least 16 AWG THWN/THHN stranded, copper, PVC-insulated, nylon jacketed. It shall be rated for 600 V to ground and 10 A maximum current. Electronic signal wire shall be at least 18 AWG for single conductor and 22 AWG for multi-conductor cable. It shall be THWN PVC-insulated, nylon jacketed.

5.3 Conductor Termination:
All terminal connections are to be screwed type with solderless crimp type connectors with insulation sleeves. Only one wire shall be terminated per connector. All cables and interconnecting wiring shall run via terminal strips. No more than two wires shall be connected to a single terminal block with the second wire being a jumper to an adjacent terminal to provide the additional connection.

6.0 Relays:

6.1 Construction:
All relays shall be industrial grade such as Allen-Bradley 700 Series or equal with DIN rail mounting.

6.2 Contact rating:
10A. continuous minimum at 48 VDC, 125 VDC, or 120 VAC.

6.3 Wiring:
The relay coil and all contact strips shall be wired to a terminal strip.

7.0 Pushbuttons and lights:
Pushbuttons, pilot lights, and switches shall be of the oil tight, heavy duty type. Pushbuttons shall have protective guards to prevent accidental actuation. Long life type de-rated lamps only shall be used.

8.0 Circuit Breakers and Fuses:

The main power to each panel section shall be isolated by means of a thermal magnetic circuit breaker while all other power circuits shall be fed from molded case circuit breakers, using a separate breaker for each circuit. Fuses and circuit breakers shall be readily accessible with proper clearance from tubing, valves, and other obstructions.

Technical Standard Document No. TS-09

Standard Title: Hydrogen Dilution Blower Specification

| Revisions | Date | Prepared By | Approved By | Purpose |

1.0 Materials:

NOTE: The following specification is provided as a standard. Proposals being submitted shall be in full compliance with this document unless an Exceltec International, hereafter referred to as EIC, client specification is provided. In this case, the client specification has precedence over areas which conflict. Supplier is to make EIC aware of all such conflicts and EIC will then instruct the supplier on how to proceed.

   1.1 Impeller:

      The impeller shall be constructed of solid FRP.

   1.2 Case:

      The casing shall be constructed of FRP laminate and use stainless steel fasteners.

   1.3 Shaft:

      The shaft shall be constructed of steel with an FRP sleeve bonded to the impeller.

2.0 Mechanical Characteristics:

Dilution blowers shall be of the radial exhauster type and shall meet the following design characteristics:

   2.1 The blower shall contain radial blade wheels of the non-overloading design. The wheels shall be statically and dynamically balanced at the factory with the lowest critical speed at least 15% below the operating speed.

   2.2 The blower shall be of the heavy duty, low speed type.

   2.3 The blower shall be connected to the motor via a flexible coupling or belt drive.

   2.4 The blower shall meet the latest edition of the Air Moving and Conditioning Association (AMCA) specifications.

   2.5 The blower housing shall contain an FRP low point drain.

   2.6 The blower shall contain shaft seals with bearings located outside of the blower housings.

   2.7 The blower shall be mounted on a steel base and connected to the system duct via ANSI 150# flanges.

*Technical Standard
Document No. TS-10*

*Standard Title: Differential
Pressure Indicator and Switch
Specification*

Revisions   Date   Prepared By   Approved By   Purpose

1.0 Applicability:
NOTE: The following specification is provided as a standard. Proposals being submitted shall be in full compliance with this document unless an Exceltec International, hereafter referred to as EIC, client specification is provided. In this case, the client specification has precedence over areas which conflict. Supplier is to make EIC aware of all such conflicts and EIC will then instruct the supplier on how to proceed.

This technical standard applies to strainer differential pressure gauges.

2.0 General Construction:
Differential pressure instruments shall be of the magnetic piston type such that the indicator is isolated from the pressure media. It shall contain multiple switches for alarm and operating functions as necessary.

3.0 Materials of Construction:
All wetted parts shall be of bronze, 316 SS, or non-metallic materials.

4.0 Mechanical Characteristics:
All differential pressure indicators shall be rated for 150 psig (9.375 bar) minimum and 150°F (65°C). They shall be capable of withstanding full system pressure from either direction. Connections shall be 1/4" NPT in line type. Gauge diameters shall be 4 1/2" with a dual scale, 0–10 PSID and 0 - 0.7 kg/cm^2.

5.0 Electrical Characteristics:
Enclosure to be NEMA 4 rated with adjustable 10 amp rated reed type switch(es).

Technical Standard Document No. TS-11

Standard Title: Solenoid Valve Specification

Revisions   Date   Prepared By   Approved By   Purpose

1.0 Application:
NOTE: The following specification is provided as a standard. Proposals being submitted shall be in full compliance with this document unless an Exceltec International, hereafter referred to as EIC, client specification is provided. In this case, the client specification has precedence over areas which conflict. Supplier is to make EIC aware of all such conflicts and EIC will then instruct the supplier on how to proceed.

This technical specification applies to air system solenoid valves which control pneumatic butterfly valve operators.

2.0 Materials:
Bronze, stainless steel or equal valve bodies with Buna-N or equal seals and discs, nylon or equal disc holder.

3.0 Mechanical Characteristics:
Solenoid valve shall be designed for 150 psig (9.375 bar) and 150°F (65°C). It shall be capable of operating against a 150 psi differential. The valve shall have 1/4" FPT system connections.

4.0 Electrical Characteristics:
Solenoid enclosure shall be NEMA 4X rated and the coil shall be continuous duty molded class F or H. The solenoid valve may be either normally open or normally closed. It shall be CSA certified and UL approved and must be capable of operating at the available system voltage and frequency.

Technical Standard Document No. TS-12

Standard Title: Transformer/Rectifier Specification for FB & SC Series Systems

Revisions	Date	Prepared By	Approved By	Purpose

1.0 Design:
NOTE: The following specification is provided as a standard. Units being provided shall be in full compliance with this document unless an Exceltec International, hereafter referred to as EIC, client specification is provided. In this case, the client specification has precedence over areas that are in conflict. Supplier is to make EIC aware of all such conflicts and EIC will then instruct the supplier on how to proceed.
1.1 Applicable standards:
1.1.1 National Electrical Code
1.1.2 ANSI C34.2
1.1.3 ANSI C57.12.01
1.1.4 ANSI C57.18
1.1.5 NEMA
1.1.6 EIC TS-019
1.1.7 ISA
1.1.8 IEEE
1.2 Units shall be suitable for indoor or outdoor use designed to NEMA 3 rating and shall be cooled by means of forced air, bottom intake, and side exhaust unless stated otherwise. These units are to be used in electrochemical duty, as defined in ANSI standard C34.2. The equipment defined in this specification will be installed in a Tropical Environment, and is expected to operate with minimum supervision and maintenance.
1.3 Ratings:
Input: Per Project
Output: Per Project
1.4 Rectification Circuit Design:
Vendor Standard
1.5 Control:
1.5.1 Constant current from 0 to 100% of rating shall be achieved by thyristor phase control using a 1 turn pot. Output shall be within (+/-) 1% over the combined effects of 50 to 100% DC output voltage at rated ambient temperature to 50°C and (+/-) 5% AC input line voltage variation.
1.5.2 Triggering circuits shall have inherent timing balance and shall be sensitive to power system distortion. Trigger shall inhibit output on the loss or reduction of an input phase.
1.5.3 Upon initiation of operation the unit shall employ a 1 – 5 second soft start to full load operation.
1.6 Environment:
Continuous operation is to be expected under maximum ambient conditions of 50°C (122°F) with 100% relative humidity. The units will be located indoor or outdoors and may be shutdown for significant periods with high humidity. The above shall be considered standard unless stated otherwise.

2.0 Construction:
- 2.1 Transformer:
  - 2.1.1 Dry type air cooled, 80°C (176°F) maximum rise.
  - 2.1.2 Insulation class H.
  - 2.1.3 Windings primary and secondary use electrolytic grade copper.
  - 2.1.4 Core — grain oriented steel, grade M6 or better.
  - 2.1.5 Coils — varnish dipped. Varnish to be Dolphs 352 or equivalent.
  - 2.1.6 All secondary connections shall be controlled in location, so that all like transformers are interchangeable, and a replacement may be installed at a later date with minimum interference.
  - 2.1.7 The transformer shall be permanently tagged with the transformer/rectifier unit serial number and date of manufacture.

    The transformer shall conform to the requirements of ANSI C57.12.01 and C57.18 for electrochemical duty.
- 2.2 Cabinet:
  - 2.2.1 A copper grounding lug of sufficient size shall be located in the main disconnect enclosure. All metallic components shall be bonded to this location.
  - 2.2.2 Transformer cubicle shall meet NEMA 3R standards, with sufficient removable access panels to permit service of any component without prior disassembly of any other components. All access panels shall be gasketed.
  - 2.2.3 A properly sized bus bar shall be provided for DC output connections as required in Section 2.3.1 of this standard.
  - 2.2.4 All ventilation openings shall be fitted with 1/2" or less mesh screens to prevent the entrance of rodents, large insects, or debris. The side exhaust opening shall be provided with a hood to direct airflow toward the floor and prevent the entrance of rain or debris.
  - 2.2.5 Control enclosure shall meet NEMA 4 standards. When a Carlon enclosure is used, the enclosure shall be mounted directly to the transformer cubicle without using the mounting feet.
  - 2.2.6 The control enclosure and disconnect box shall be of sufficient size as to facilitate service. Provision shall be made for the purchaser's cable/conduit entrance from either the top or bottom. All components shall be attached in a manner in which they can be removed without removing the backpan.
  - 2.2.7 Wire and cable entries between the control enclosure and transformer cubicle shall be sealed.
  - 2.2.8 EIC paint specification TS-019 shall apply to the main rectifier enclosure.

2.2.9 All fasteners 1/4" and larger shall be 316 stainless steel. All fasteners smaller than 1/4" shall be 18–8 stainless steel. A minimum of 1 thread shall penetrate from the nut.

2.3 Rectifier Section:

2.3.1 Conductors shall be electrolytic grade copper of at least one square inch per 1000 Amps in each path for a forced air unit.

2.3.2 Power semiconductors shall have blocking voltage ratings of a minimum of three times applied voltage. Sufficient cooling shall be provided to keep the maximum case temperature at least 15°C (59°F) below the maximum rating for the device at rated current and at rated maximum coolant temperature.

2.3.3 Stud, module, and puck type SCRs shall be acceptable. Module type SCRs shall be environmentally protected.

2.3.4 Rectifier outputs exceeding 500 Amps DC rating shall have secondary power semiconductors individually fused with current limiting fuses of the appropriate rating to isolate a faulted path without further damage to the unit.

2.4 Cooling System:

2.4.1 Cooling of the power components is by forced air unless otherwise stated. Power semiconductors shall be mounted on properly sized aluminum heat sinks. Heat sinks shall be anodized, irradiated, or thoroughly coated with protective varnish.

2.5 Wiring:

2.5.1 All control wiring shall be:
- Minimum 18 awg.
- Industrial quality.
- Rated at 90°C (196°F) or better.
- Rated for 600 V.
- Conductors shall be fine stranded copper.
- Insulation shall be PVC.
- Wire shall be UL listed.
- Thyristor gate and cathode leads shall be manufacturer's wire gauge or larger.

2.5.2 Terminations shall either be by full fork tongue compression type terminals which have insulated compression sleeves to grip the wire, industrial grade clamp type terminal strips, or pressure terminals in which case the stripped wire may be inserted directly under the pressure plate. Alternatives may be utilized if approved by an EIC engineer.

2.5.3 All wiring shall be labeled with Brady tape type wire number labels with a consistent numbering scheme. Wire numbers shall be resistant to oil, water, and most solvents.

Wire numbers shall be suitable for use in a 50°C ambient temperature. All wire numbers shall be shown on the electrical drawings. Wire numbers for a series sequence of circuit elements shall increase by one for each element passed through. Wire numbers shall be readable from the top down or from left to right.

2.5.4 No more than two wires shall be landed on any termination point. Only one wire or cable shall be crimped into any lug or terminal. Butt splices, wire nuts, and other methods of wire splicing are prohibited except where a device is provided with wires installed. In this case, every effort shall be made to connect the wires to a terminal strip before considering the use of a wire splice. Ground lugs shall be permitted to contain more than two wire terminations.

2.5.5 Wire ducts shall be used to organize and route control wiring. Proper routing shall provide separation between AC control and signal wiring. Wire ducts shall be sized so that the enclosed wiring consumes no more than 80% of the total available space. Wire ducts shall maintain the same height in any plane.

2.5.6 All wiring not routed through wire ducts shall be secured using epoxy affixed type ty-wrap bases. White adhesive backed ty-wrap bases are not acceptable.

2.5.7 Power wiring shall be of a size and insulation level appropriate to its application.

2.5.8 External pilot devices shall be identified with engraved plastic tags (Gravopoly) and attached by stainless steel rivets or approved adhesive. Tags shall be black with white lettering and shall correspond with all drawings.

2.6 Terminal Blocks:
2.6.1 All wiring for external field connections shall be terminated on a minimum number of conveniently located terminal blocks.

2.6.2 Terminal blocks shall be mounted so that sufficient clearance shall be provided for wire insertion, removal, and for the purchaser's wiring connection. A three-inch clearance shall be maintained adjacent to terminal blocks for wire routing.

2.6.3 Terminal blocks shall be of adequate size and shall be designed to receive purchaser's incoming control cables.

2.7 Controls:
2.7.1 General service auxiliary control relays shall be a plug in sealed design type IDEC or equal. Only one contact of each form C pair is to be utilized, except for the customer system alarm and rectifier operation status where both

NO and NC contacts are permitted. The vendor may utilize PLC logic if approved by EIC.

 2.7.2 Dry circuits (current < 50 mA, voltage < 30 V) must be switched by relays rated for the application, IDEC type RY4S-UL120 VAC, or approved equal.

 2.7.3 Push buttons, switches, and lights shall be industrial quality, oil tight construction suitable for NEMA 4X application.

 2.7.4 Any edge connecting circuit boards must have a system for mechanically securing the boards and have a corrosion resistant plating system for the connecting edge, such as Ni or Au.

 2.7.5 All electronic boards shall be completely tropicalized before installation. If any board adjustments are made during testing these areas should be touched up at the conclusion of the test.

 2.7.6 Proper protection (i.e., fuses, circuit breakers) shall be provided for all equipment as dictated by the National Electric Code.

2.8 Tropicalization:

All electrical equipment, internal bus bars, and their enclosures shall be tropicalized. Secondary wiring, coils, and other insulation that are not fungus resistant, shall have a fungus-resistant coating applied, except where such coatings would interfere with the proper operation. In such cases, the part shall be inherently fungus resistant. Dolph's Synthite AC-279–7s Clear Air Drying Anti-fungal Varnish or EIC approved equal shall be used. Care shall be taken to prevent excessive overspray, runs, and drips on painted surfaces.

2.9 Motors:

All motors shall be totally enclosed fan cooled (TEFC) rated for the design ambient conditions. Insulation shall be Class F Continuous Duty.

3.0 Protection and Metering:

 3.1 Design:

 Vendor shall be responsible for proper unit protection coordination.

 3.2 Protection:

  3.2.1 Circuit protection shall be supplied as follows:

   3.2.1.1 Fuses or an incoming power circuit breaker shall be a manually operated type. Design shall be sized to correctly protect the transformer/rectifier.

   3.2.1.2 Auxiliary power shall be protected by fuses or circuit breakers which are manually operated type. Individual fuses or circuit breakers should be provided for the 110/120 VAC control power.

   3.2.1.3 Circuit breakers shall be of the indicating type providing ON, OFF, and TRIPPED positions of the operating handle. Multiple breakers shall be designed so that an overload on one pole shall open all poles.

3.3 Alarms and Shutdowns:
   3.3.1 Each of the following conditions shall be indicated locally and cause shutdown. These indications shall be latching indications, which are reset by an acknowledge pushbutton. A shutdown condition causes the thyristor triggering to be inhibited, and/or the primary AC circuit contactor to be opened. A test push-button should also be provided for periodic testing of all alarm lamp indicators. All alarm and shutdown circuits shall be designed using Fail Safe logic.
      3.3.1.1 Semiconductor fuse failure shutdown. Visual indication of individual failed devices shall be provided.
      3.3.1.2 Semiconductor over-temperature shutdown (one sensor for each semiconductor assembly) and transformer over-temperature shutdown. Sealed thermo-switches shall be utilized.
      3.3.1.3 DC Over-voltage shutdown, set at 105% of rated output.
      3.3.1.4 DC Over-current shutdown, set at 105% of rated output.
      3.3.1.5 Motor(s) failure overload trip.
   3.3.2 Remote indication shall be provided by means of a Form C isolated contact for each of the following functions.
      3.3.2.1 Rectifier Operational indication.
      3.3.2.2 Tripped - Fault relay contacts.

3.4 Metering:
The following meters shall be provided as a minimum on each unit. Meters shall have an accuracy of +/- 2% of full scale suitable for NEMA 4 application. All required instrument transformers shall be supplied.
   3.4.1 DC Ammeter
   3.4.2 DC Voltmeter
   3.4.3 Elapsed running time meter

3.5 Panel Controls:
The following panel controls shall be mounted on the control cabinet of each individual unit.
   3.5.1 Auxiliary control circuit (circuit breaker or fuses)
   3.5.2 Current control potentiometer

4.0 Quality Control and Testing:
  4.1 Documents:
     4.1.1 Project schedule.
         Within two weeks of receipt of order, the vendor shall provide an estimated project schedule showing the anticipated dates of major events in the project.

4.1.2 Approval drawings.
The following drawings are required for customer approval, before start of construction. For the purpose of scheduling, assume 2 weeks from submittal for drawing approval. Drawings shall be submitted by either e-mail or magnetic media.

4.1.2.1 Mechanical outline drawings showing detailed cabinet outlines and dimensions, power and control entry locations and dimensions, service access, control locations, panel layout, and provisions for equipment handling. Mechanical outline drawings required for both the rectifier and control enclosures.

4.1.2.2 Electrical schematic drawings showing the close-up wiring detailing all protective circuits and rectifier control logic.

4.1.3 Final drawings and documents.
One D size drawing with a minimum 1/8" text shall be submitted as well as a magnetic disk for reproducing additional copies for insertion into manuals.

4.1.3.1 Mechanical outline and installation footprint drawing.

4.1.3.2 Major component layout and service access drawing.

4.1.3.3 Internal schematic diagram(s) showing the power circuit, protection circuits, controls, customer interface, and metering circuits. All wire numbers shall be shown. Drawing formats and symbols shall comply with ISA Standards and IEEE Standards for electrical circuits.

4.1.3.4 Field installation wiring diagram, showing all required field termination, terminal locations, terminal numbers, and wire sizes if not a part of 4.1.3.3.

4.1.3.5 Electronic control schematic and layout drawing if not part of 4.1.3.3.

4.1.3.6 Instruction manuals. One copy shall be supplied.

4.1.3.7 Bill of materials showing component specifications of all major components such as power semiconductors, protective relays, connectors, fuses, etc.

4.1.3.8 Vendor catalogue cuts shall be accumulated, bound in a 3-ring notebook, and provided to EIC as part of the project data. These data should include manufacturer's part numbers and full descriptions of each component to allow ease of UL inspection and approval.

4.1.4 Installation, Operating, and Maintenance Manual. The specified number of copies shall be submitted after the draft copy is approved. Manuals should contain the following as a minimum.
4.1.4.1 Equipment description, layout, and operation.
4.1.4.2 Maintenance instructions.
4.1.4.3 Electronic controls and testing.
4.1.4.4 Complete Bill of Materials showing manufacturers and ratings. Supplier's correct part numbers must be included.
4.1.4.5 Complete detailed test reports, as described elsewhere.
Certified As Built B size drawings as described in Section 4.1.3 shall be supplied as part of this manual.

4.2 Inspection:
  4.2.1 In-process inspection.
  EIC reserves the right to inspect progress on this project at any time.
  4.2.2 Final inspection.
    Final inspection shall be witnessed by EIC, or their representative, at full load of one unit. The vendor shall have completed a full load heat run on all units prior to this witness test. Vendor shall notify EIC at least 2 weeks prior to that point of the test schedule. This inspection may be waived by EIC.
  4.2.3 Test reports.
    Test procedures and report forms shall be submitted in detail to EIC for approval prior to testing new or revised designs of transformers and rectifiers. At least 2 weeks shall be allowed for this approval process.
  4.2.4 Final test reports shall be submitted to EIC after inspection and test.
  4.2.5 All test equipment shall be calibrated at established intervals. All instrument serial numbers shall be recorded on the test reports.

4.3 Transformer tests:
  Transformer tests shall be performed in accordance with ANSI standards as per the following items. Test reports on each transformer shall be traceable to the unit in which it is installed. A copy of the transformer test report shall be included in the instruction manual for that unit.
  4.3.1 Unit tests:
    Each transformer shall be given the following tests prior to installation in the rectifier units. These tests are to be noted on the test report and shall be traceable to the transformer/rectifier serial number.

4.3.1.1 Dielectric test.

4.3.1.2 Noise level shall not exceed 85 dBA at 1 meter.

4.4 Electronics:

4.4.1 Board traceability must be provided on the unit Bill of Materials.

4.4.2 Calibration data shall be provided for any calibrated control on the unit test data sheet.

4.5 Unit Tests:

The following tests shall be conducted in accordance with ANSI standard C34.2. Copies of all test reports shall be included in the instruction manual.

4.5.1 Type tests:

These tests shall be conducted on one unit of each design, prior to the witness test and shall be documented on the general test report.

4.5.1.1 Unit power losses at 25, 50, 75, and 100% of rated output current.

4.5.2 Unit tests:

These tests shall be conducted on each unit, and documented on the unit test report.

4.5.2.1 Dielectric tests on each independent circuit section, i.e., high voltage, DC output, AC control, protection circuits, etc.

4.5.2.2 Functional tests on all protective and control circuits.

4.5.2.3 Full load test. Semiconductor case temperatures are to be probed and recorded after 1 hour of operation.

4.5.2.4 Full current and full voltage tests per ANSI standard C34.2.

4.5.2.5 Temperature rise at full load. Temperature recordings shall be taken of transformer windings, semiconductor cases, magnetic cores, internal ambient, and external ambient until all temperatures stabilize. Temperatures are to be recorded using a chart recorder. Stability is reached when no temperature rise is recorded in a 1-hour period.

Technical Standard Document No. TS-13

Standard Title: Pressure Indicator Specification

Revisions   Date   Prepared By   Approved By   Purpose

1.0 Applicability:
NOTE: The following specification is provided as a standard. Proposals being submitted shall be in full compliance with this document unless an Exceltec International, hereafter referred to as EIC, client specification is provided. In this case, the client specification has precedence over areas which conflict. Supplier is to make EIC aware of all such conflicts and EIC will then instruct the supplier on how to proceed.

This technical standard is applicable to all pressure indicators throughout the *SANILEC* system.

2.0 General:
Pressure instruments shall be direct reading type with a range of 1.5 times the maximum normal operating pressure. Pressure gauges shall have dual scales showing PSI and kg/cm^2.

3.0 Materials of Construction:
Pressure gauges shall have stainless steel or equal bourdon tubes and movements. Gauges shall have stainless steel, PVC, or equal cases and acrylic, laminated safety glass, or equal windows.

4.0 Mechanical Characteristics:
Pressure instruments shall be 4" to 4 1/2" with an MPT service connection. They shall feature an open front/blow out disc construction and shall be accurate to within 2% of full scale or better for temperatures between 0–120°F (–18–38°C).

Technical Standard Document No. TS-14

Standard Title: Gauge Guard Specification

Revisions   Date   Prepared By   Approved By   Purpose

1.0. Applicability:
NOTE: The following specification is provided as a standard. Proposals being submitted shall be in full compliance with this document unless an Exceltec International, hereafter referred to as EIC, client specification is provided. In this case, the client specification has precedence over areas which conflict. Supplier is to make EIC aware of all such conflicts and EIC will then instruct the supplier on how to proceed.

This technical standard is applicable to all gauge guards (diaphragm seals) used in the *SANILEC* system.

2.0. Materials of Construction:
The housing of the gauge guard shall be constructed of PVC or equal, the diaphragm shall be constructed of Teflon, Viton, or equal and the O-ring seals shall be Buna-N, Viton, or equal.

3.0 Mechanical Characteristics:
Gauge guards shall be oil filled with a 1/2" process connection and a 1/4" or 1/2" instrument connection as specified. It shall be of two piece construction with stainless steel fasteners.

Technical Standard Document No. TS-15

Standard Title: Flow Orifice Specification

Revisions   Date   Prepared By   Approved By   Purpose

1.0. Applicability:
   NOTE: The following specification is provided as a standard. Proposals being submitted shall be in full compliance with this document unless an Exceltec International, hereafter referred to as EIC, client specification is provided. In this case, the client specification has precedence over areas which conflict. Supplier is to make EIC aware of all such conflicts and EIC will then instruct the supplier on how to proceed.
2.0. General:
   Flow control orifices shall be of the multiple flexible insert type to provide a steady flow of +/- 10% at variable pressures greater than 25 psig.
3.0. Materials of Construction:
   Flow control orifices shall be constructed entirely of non-metallic materials for corrosion resistance. The body shall be constructed of PVC or equal and the inserts shall be constructed of Buna-N or equal.
4.0. Mechanical Characteristics:
   Flow control valve shall be rated for 120 psig (7.5 bar) and 100°F (38°C). It shall have either plain or flanged connections.

*Technical Standard
Document No. TS-16*

*Standard Title: Rotary Sensor Type
Flow Indicator Specification*

Revisions   Date   Prepared By   Approved By   Purpose

1.0 Applicability:
NOTE: The following specification is provided as a standard. Proposals being submitted shall be in full compliance with this document unless an Exceltec International, hereafter referred to as EIC, client specification is provided. In this case, the client specification has precedence over areas which conflict. Supplier is to make EIC aware of all such conflicts and EIC will then instruct the supplier on how to proceed.

This technical standard defines the minimum technical requirements for cell inlet flow meters.

2.0 General:
Flow indicators shall consist of a rotor type sensor, an installation tap, and a remote flow indicating unit.

3.0 Materials of Construction:
The sensor shall be constructed of polypropylene or equal, with titanium or equal shaft. The tap shall be constructed of PVC or equal.

4.0 Mechanical Characteristics:
The sensor and tap assembly shall be capable of withstanding 100 psig (6.25 bar) @ 120°F (50°C). The sensor shall be mounted in a wet tap which allows removal of the sensor without securing the system.

5.0 Electrical Characteristics:
The flow sensor shall produce a variable frequency output of at least 1 volt and be accurate to 2% of full range. The flow indicating unit shall have a local alarm and a remote alarm relay with contacts rated at 3 A at 240 V or better. The indicating unit shall have an accuracy of 2% of full scale.

Technical Standard Document No. TS-16A

Standard Title: Vane Type Flow Indicator Specification

Revisions   Date   Prepared By   Approved By   Purpose

1.0 Applicability:
NOTE: The following specification is provided as a standard. Proposals being submitted shall be in full compliance with this document unless an Exceltec International, hereafter referred to as EIC, client specification is provided. In this case, the client specification has precedence over areas which conflict. Supplier is to make EIC aware of all such conflicts and EIC will then instruct the supplier on how to proceed.

This technical standard defines the minimum technical requirements for in-line vane type cell inlet flow meters.

2.0 General:
Flow indicators shall consist of an in-line flow indicator equipped with a transmitter (if applicable) to provide alarm and control functions.

3.0 Materials of Construction:
The housing shall be constructed of PVC or equal, with titanium, Hastelloy-C or equal internal wetted parts and Viton or equal seals.

4.0 Mechanical Characteristics:
The flow meter shall have a maximum operating pressure of 100 psig (6.25 bar) at 100°F (38°C). The flow meter shall be equipped with ANSI 150# flanges for installation into the system. The flow meter shall create a pressure drop of not more than 2.5 PSI.

5.0 Electrical Characteristics:
The flow meter control box shall be NEMA 4 rated and shall be UL listed. It shall contain SPDT switches rated at 15A, at 480 VAC which may be field adjusted. The unit shall be accurate to within 2% for over the full range of the monitor. If included, the transmitter shall be a two-wire model with a 4–20 mA output.

Technical Standard Document No. TS-17

Standard Title: Low Flow Switch Specification

Revisions   Date   Prepared By   Approved By   Purpose

1.0 Applicability:
NOTE: The following specification is provided as a standard. Proposals being submitted shall be in full compliance with this document unless an Exceltec International, hereafter referred to as EIC, client specification is provided. In this case, the client specification has precedence over areas which conflict. Supplier is to make EIC aware of all such conflicts and EIC will then instruct the supplier on how to proceed.

This specification applies to low flow switches mounted on the electrolyzer cell inlet or outlet.

2.0 Materials of Construction:
Switch body shall be constructed of corrosion-resistant non-metallic material such as Ryton, Kynar, Teflon, PVC, etc. Seals shall be constructed of Viton and all other wetted parts shall be non-metallic where practicable or Hastelloy-C or titanium where metal is required.

3.0 Type:
Switches shall be of the paddle type and shall operate independently of system pressure or temperature.

4.0 Mechanical Characteristics:
Flow switches shall be constructed with a 1" MPT connection for entering the system. The switch shall be rated for a minimum working pressure of 60 psig (3.75 bar) and minimum working temperature of 150°F (65°C).

5.0 Electrical Characteristics:
The switch shall be SPDT with a contact rating of 10 Amps at 120 VAC.

Technical Standard Document No. TS-19

Standard Title: Powder Coating Specification

Revisions   Date   Prepared By   Approved By   Purpose

1.0 Scope:
  1.1 This specification covers the painting of cabinets and related panel structures which are not otherwise specified on engineering drawings or by individual equipment specifications.
  1.2 Painting shall include surface preparation, under coatings, topcoats, inspection, and touch up.
2.0 Reference Standards:
  SSPC Systems and Specifications
  SSPC-PA-1 Shop, Field, and Maintenance Painting
  SSPC-PA-2 Measurement of Dry Paint Thickness with Magnetic Gauges
3.0 Contractor's Responsibility:
  3.1 The contractor shall furnish all materials, tools, equipment, labor, and supervision to execute and fulfill the intent of this specification.
  3.2 The contractor shall be responsible for preparation of all surfaces to be painted in compliance with the surface preparation procedure as described in Paragraph 5.0 of this specification and that of the paint manufacturer, to the satisfaction of the purchaser's representative.
  3.3 The contractor shall be responsible for maintaining the painting area at the proper temperature and humidity with respect to those recommended by the paint manufacturer and a state of cleanliness acceptable for performance of top quality workmanship.
4.0 Materials:
  4.1 The paint as specified herein shall be maintained in strict accord with manufacturer's specifications.
  4.2 No other paint or painting system than that which is specified shall be acceptable, without prior approval by the purchaser.
5.0 Surface Preparation:
  5.1 All welded areas shall be given special attention for removal of welding flux in crevices. Weld spatter, slivers, sharp edges, laminations, and underlying mill scale not removed during fabrication shall be removed by best mechanical means and edges smoothed prior to cleaning.
  5.2 All surfaces to be painted shall be thoroughly cleaned using the paint manufacturer's recommendations for removal of oil or grease.
  5.3 Wash using inline automatic washer utilizing paint manufacturer's recommended alkaline base cleaner.
  5.4 Rinse thoroughly with fresh water.
  5.5 Apply iron phosphate solution to the parts for preparation of the surface.
  5.6 Remove applied phosphate with a fresh water rinse.

5.7 Apply manufacturer's recommended sealer over the surface to assure maximum adhesion and protection.
6.0 Coating Application:
   6.1 Surfaces to be painted shall be sealed immediately after cleaning. No paint shall be applied to sealed surfaces which have been allowed to set overnight or for a maximum 8-hour shift, or which have been subjected to inclement weather.
   6.2 Paint shall be applied by the electrostatic method, of a uniform thickness and finish as specified herein.
   6.3 Successive paint applications, if required, shall specifically follow the recommendations of the paint manufacturer.
7.0 Protective Coating System for Exterior Surfaces:
   7.1 The coating system shall consist of an inorganic phosphate coat and a weather resistant polyester or urethane topcoat. Total dry film thickness shall be 3–6 mils.
   7.2 The coating application shall consist of the following three coat system:
   Sealer Coat: To be wetted over the entire surface to assure full coverage.
   Intermediate Coat: To be 0.5–1 mil d.f.t. organic phosphate.
   Top Coat: To be 3–5 mils d.f.t.; polyester, Tiger Drylac #5015 blue, fine texture.
   7.3 Cure to be 350°F–425°F for 20–30 minutes.
8.0 Inspection:
   8.1 All work and materials supplied under this specification shall be subject to timely inspection by the purchaser or his authorized representative. The contractor shall correct such work or replace such material as is found defective under this specification.
   8.2 In-process inspections will be performed in accordance with the EIC inspection and test plan (if provided). EIC or their authorized representative shall be notified 5 days in advance of any inspections or hold points outlined in the inspection and test plan.
   8.3 EIC or their authorized representative shall be notified 2 days prior to order completion for final inspection of the completed product.
   8.4 Paint dry film thickness of intermediate and finish coats shall be checked per specification requirements using a magnetic pull-off type gauge. Alternative paint thickness gauges are to be approved by the purchaser and checked prior to inspection by the purchaser's representative.
   8.5 Magnetic pull-off type gauges used for measurement of coating thickness shall be calibrated with calibration standards traceable to NIST.

8.6 A test and inspection report shall be provided for all coating work performed. This report shall include the following minimum data:
- Date and time coating was applied
- Coating system applied
- Coating color
- Required and measured coating thickness
- Date coating thickness was measured

8.7 A certificate of compliance, signed and dated by the contractor's authorized representative, shall be furnished for each shipment of material or services supplied. This certification shall state that all items furnished in the shipment are in full compliance with all purchase order and specification requirements.

9.0 Paint Manufacturers and Products:

9.1 Following are approved paint manufacturers and products for exterior surfaces:

Manufacturer	Products[a]
sealer	inter. coat topcoat
Tiger	RAL 5015
Drylac	Fine Texture

[a] Note: Other paint manufacturers and products may be submitted to EIC for approval.

*Technical Standard
Document No. TS-21*

*Standard Title: Electrolyzer
System Specification*

Revisions	Date	Prepared By	Approved By	Purpose

NOTE: The following specification is provided as a standard. Systems being provided shall be in full compliance with this document unless an Exceltec International, hereafter referred to as EIC, client specification is provided. In this case, the client specification has precedence over areas which conflict.

1.0 Electrolyzer Module Construction:
   1.1 Scope:
   This specification covers the electrolytic cells, interconnecting piping and valves, bus-work, and all other equipment that is necessary to make the production streams complete, contributing to smooth reliable plant operation.
   1.2 Steel Skid:
      1.2.1 The design of this system ensures reliable operation in a chemical and corrosive environment. The system skid shall be constructed of A36 structural steel.
      1.2.2 All steel shall be coated in accordance with EIC technical standard TS-025, latest revision.
   1.3 Electrolytic Cells:
      1.3.1 The electrolytic cell assemblies shall be supplied with softened water and concentrated brine. This stream shall be pressurized to a minimum of 45 PSI (three (3) bar g). The sodium hypochlorite solution produced shall be sent to the storage tank with the entrained hydrogen gas for hydrogen removal at the storage tank.
      1.3.2 The flow of dilute brine through the electrolytic cell assemblies shall be kept at a constant rate sufficient for addition to the storage tank without the need for transfer pumps.
      1.3.3 All cell assemblies shall be a monopolar plate type designed and constructed to enable easy removal and replacement of cell electrodes with due regard to the need for inspection and maintenance.
      1.3.4 The anodes and cathodes shall be spaced uniformly with a gap of 0.040 inches to provide for optimum cell efficiency. Internal spacers shall be placed strategically over the electrode surface to maintain a uniform gap. Spacer material shall be Teflon (polytetrafluroethylene) to withstand the corrosive operating environment.
      1.3.5 Cell anodes shall be expanded titanium metal with a precious metal oxide coating and operated at a current density of less than 1,600 amps/m^2. They shall be dimensionally stable type and shall preserve their shape and voltage characteristics under severe electrolytic conditions.
      1.3.6 Cell assembly cathodes shall be constructed of expanded Hastelloy C-276 material to offer superior corrosion resis-

tance to seawater and sodium hypochlorite while improving power consumption by 10% comparative to the use of titanium cathodes.

1.3.7 The cell housing shall be injection molded polypropylene with ultraviolet stabilizers and integrally molded flanges to ensure positive sealing.

1.3.8 The maximum applied voltage differential (per cell) shall not exceed 4.2 V DC during operation. A feed brine temperature of 25°C with a chloride content of 19,000 ppm should be used for this basis.

1.3.9 All internal cell assembly fasteners shall be commercially pure titanium. All external cell assembly fasteners shall be 316 stainless steel, copper, brass, or silicone bronze.

1.3.10 All cell assembly gaskets or seals shall be either Viton or silicone rubber.

1.4 Bus Bar:
Interconnecting cell assembly bus bar shall be constructed from 99.8% pure electrolytic grade copper.

1.5 Piping:
All piping shall be Schedule 80 PVC with ultraviolet stabilizers, Type 1, Grade 1. All PVC Schedule 80 pipe shall conform to ASTM D-1784 and D-1785 standards. Schedule 80 socket fittings shall meet ASTM D-2467 while threaded fitting shall conform to ASTM D-2464. Both pipe and fittings shall be the product of one manufacturer.

1.6 Fasteners:
All fasteners for assembly of structural or piping components and for mounting of components shall be 316 stainless steel.

1.7 Valves:
Valves shall be provided as follows:
  1.7.1 Ball Valves:
    Shall be Schedule 80 PVC, True Union type with Viton O-ring seals and Teflon seats.
  1.7.2 Butterfly Valves:
    Shall be Schedule 80 PVC with 316 SS shafts and EPDM seats and shaft seals. Fasteners shall be 316 stainless steel.
  1.7.3 Globe Valves:
    Globe valves shall be schedule 80 PVC socket or threaded with EPDM seals and seats.
  1.7.4 Actuators:
    As applicable, ball or butterfly valves shall be equipped with pneumatic operated or electric actuators. Design shall be Fail-Spring to Close. Actuator housings shall be NEMA 4 construction and shall contain two (2) limit switches to be used for remote indication.

1.8 Instruments:
The electrolyzer module shall contain, as a minimum, the following instruments:

1.8.1 Inlet Flow Meter:

Type:	Variable area
Accuracy:	+/- 5% of full range
Repeatability:	+/- 2% of full range
Construction:	Bronze and glass
Indication:	Bronze float

1.8.2 Pressure Relief Valve:

Type:	Schedule 80 PVC; Type 1, Grade 1; Angle Pattern Design
Seals:	Viton
Shaft:	Teflon
Fasteners:	316 Stainless Steel
Spring:	Stainless Steel

1.8.3 Pressure Gauge (Inlet and Outlet):

Type:	Process Gauge — Liquid Filled
Case Material:	Glass Reinforced Thermoplastic or S.S.
Window Material:	Acrylic
Dial Size:	2"
Accuracy:	1% of Full Range
Bourdon Tube:	316 Stainless Steel
Scale:	Dual Scale — "PSI"
Connection:	Lower Mount — 1/4" NPT
Safety Feature:	Solid Front — Blow Out Back
Gauge Guard:	PVC Type — 1/2" NPT Process connection 1/4" NPT Instrument connection

1.8.4 Low Water Flow Switch:

Type:	Paddle Actuator
Construction:	Brass body, Buna-N, Red Brass, SS, Phosphor Bronze, & PVC wetted parts
Switch Contact:	SPDT, 15 A at 240 V A.C.

1.8.5 Inlet Temperature Gauge (Note: As required):

Type:	Bi-metal Thermometer — Liquid Filled
Case Material:	Glass Reinforced Thermoplastic
Window Material:	Acrylic
Dial Size:	5"
Accuracy:	1% of Full Span
Scale:	"deg C"
Connection:	Every angle - 1/2" NPT

1.9 Rectifier:
    1.9.1 Rectifiers shall be manufactured in accordance with EIC technical standard TS-012, Transformer/Rectifier Specification For FB/SC Series Systems, latest revision.
    1.9.2 Rectifier enclosures shall be coated in accordance the EIC technical standard TS-019, Powder Coating Specification, latest revision.

2.0 Inspection and Testing:
    2.1 Inspection:
        2.1.1 In-process inspection shall be performed where required on the job router.
        2.1.2 Final visual and dimensional inspection of each system shall be performed by the EIC QC inspector.
        2.1.3 Inspection report forms shall be completed to properly document all inspections and tests.
    2.2 System Tests:
        2.2.1 Hydrostatic testing of the system at 55 PSI for 1 h.
        2.2.2 Functional testing of all protective and control circuits.
        2.2.3 The system shall be operated at full voltage and full current.
        2.2.4 All system tests and documentation shall be verified by the QC inspector.
        2.2.5 Test report forms shall be completed to properly document all inspections and tests.

*Technical Standard
Document No. TS-22*

*Standard Title: Ball Valve
Specification*

| Revisions | Date | Prepared By | Approved By | Purpose |

1.0 Applicability
NOTE: The following specification is provided as a standard. Proposals being submitted shall be in full compliance with this document unless an Exceltec International, hereafter referred to as EIC, client specification is provided. In this case, the client specification has precedence over areas which conflict. Supplier is to make EIC aware of all such conflicts and EIC will then instruct the supplier on how to proceed.

This Technical standard applies to all size manual ball valves within the system.

2.0. Materials
The ball, body, stem, seal carrier and retaining ring shall be PVC, CPVC or equal. The seats shall be Teflon or equal, and the seals shall be Viton, EPDM or equal.

3.0. Style
Standard valves shall be True Union Ball Valves.

4.0. Design
Valves shall be designed for 100 PSIG @ maximum system operating temperature. Valves shall have an external adjustment to compensate for seat wear and shall have double block design to stop flow in either direction. The valve shall have a full port to minimize flow restrictions.

Technical Standard Document No. TS-23

Standard Title: Butterfly Valve Specification

Revisions   Date   Prepared By   Approved By   Purpose

1.0 Applicability:
NOTE: The following specification is provided as a standard. Proposals being submitted shall be in full compliance with this document unless an Exceltec International, hereafter referred to as EIC, client specification is provided. In this case, the client specification has precedence over areas which conflict. Supplier is to make EIC aware of all such conflicts and EIC will then instruct the supplier on how to proceed.

This technical specification applies to all sizes of manually operated butterfly valves.

2.0 Materials of Construction:
Butterfly valve bodies shall be constructed of PVC or equal, shaft/disc shall be constructed of PVC or equal, shaft/stem shall be constructed of titanium, 316 SS or equal, and seats and seals shall be constructed of Viton, EDPM, or equal.

3.0 Mechanical Design Characteristics:
Butterfly valves shall be designed for 100 psig at maximum system operating temperature.

4.0 Required Features:
1. Valves shall have a dry type shaft seal to prevent liquid from contacting the shaft.
2. Valve shall contain Teflon or equal bearings to ease operation and extend service life.
3. Valve shall contain a wafer type disc to minimize flow restrictions.

Technical Standard Document No. TS-24

Standard Title: Pressure Relief Valve Specification

Revisions   Date   Prepared By   Approved By   Purpose

1.0 Applicability:
NOTE: The following specification is provided as a standard. Proposals being submitted shall be in full compliance with this document unless an Exceltec International, hereafter referred to as EIC, client specification is provided. In this case, the client specification has precedence over areas which conflict. Supplier is to make EIC aware of all such conflicts and EIC will then instruct the supplier on how to proceed.

This technical specification is applicable to pressure relief valves designed to protect electrolytic cells and surrounding piping from overpressure.

2.0 General:
Relief valves shall be self-actuated angle type relief valves with a set point of 40 psig (2.5 bar).

3.0 Materials of Construction:
Relief valve body shall be constructed of PVC or equal, shaft shall be constructed of Teflon or equal, seats and seals shall be constructed of Viton or equal, and assembly hardware shall be constructed of 316 SS.

4.0 Mechanical Characteristics:
The relief valve shall have a maximum accumulation of 40% and blow-down of 25%. Relief valve shall be capable of relieving the entire system flow rate when fully open. If this requirement is not met, additional relief valves may be installed in parallel to satisfy this requirement. Relief valves shall have Female National Pipe Thread (FNPT) connections or be flanged for ease of installation.

5.0 Required Features:
1. Relief valve shall be equipped with a telltale drain to indicate seal failure.
2. Relief valves shall be equipped with multiple seals to permit continued operation until the next scheduled shutdown.
3. Relief valves shall be factory preset with a provision for owner adjustment should it become necessary.

Technical Standard Document No. TS-26

Standard Title: Induction Motor Specification

| Revisions | Date | Prepared By | Approved By | Purpose |

1.0 Applicability:
NOTE: The following specification is provided as a standard. Proposals being submitted shall be in full compliance with this document unless an Exceltec International, hereafter referred to as EIC, client specification is provided. In this case, the client specification has precedence over areas which conflict. Supplier is to make EIC aware of all such conflicts and EIC will then instruct the supplier on how to proceed.

This technical specification defines the minimum technical specification for induction motors of 1 HP or greater.

2.0 General:
Motors shall conform to the requirements of NEMA/ANSI MG 1–1987, Rev. 1 subject to the following additions or exceptions with the applicable paragraph in parenthesis.
1. Motors shall be NEMA frame suitable for chemical and other severe duty applications commonly referred to as severe duty motors. Motors shall be of the totally enclosed type.
2. The vendor shall provide a rotatable, diagonally split, gasketed main terminal housing (not required for explosion-proof motors). Terminal housings shall be provided with threaded conduit hubs. (MG 1–4.02)
3. For explosion-proof motors, the class, division, and group shall be indicated on the nameplate, together with the agency authorized to certify the equipment. (MG 1–10.38)

Technical Standard Document No. TS-31.1

Standard Title: Transformer Specification

Revisions   Date   Prepared By   Approved By   Purpose

1.0 Scope:

This specification describes the requirements for a group of rectifier transformers, which are to be built into a set of thyristor supplies for electrochemical duty. These units are to be fed by an industrial quality power source and will deliver power on an intermittent basis.

2.0 Design Requirements:
   2.1 Configuration — Vendor Standard.
      2.1.1 Primary to be Delta connected.
      2.1.2 Secondaries to be Wye connected.
   2.2 Cooling shall be ambient convection — ambient average 40°C.
   2.3 Input power — The input power available is 460 Volts +/-5%, 60 Hz. The source has a fault capacity of 40 KVA.
   2.4 Output Voltage — to meet 50 Volts DC per attached circuit design.
   2.5 Impedance — Total commutating impedance shall be in the range of 2%–5% on the rated KVA base. Every effort shall be made, consistent with good design practice, to minimize the difference between secondary coil impedance.
   2.6 Loss — total loss shall be less than 9 KW at full rating.

3.0 Construction:
   3.1 Dry type convection air cooled.
   3.2 Insulation — class H per NEMA Spec. MK-35C.
   3.3 80°C maximum rise.
   3.4 Windings — primary and secondary copper.
   3.5 Core — grain oriented steel, grade M6 or better.
   3.6 Coils — electrolytic grade copper.
   3.7 Varnish—to meet MIL specification M-32102-CE rated to 215°C or better.
   3.8 Transformer — over temperature. One sensor each phase at 150°C for trip.
   3.9 All secondary connections shall be controlled in location.
   3.10 The transformer shall conform to the requirements of ANSI standards C57.12.01 and C57.18.

4.0 Transformer Tests:

These tests shall be done in accordance with ANSI C57.12.01 and C57.18. Test reports shall be supplied for each unit.
   4.1 Type tests. The following tests shall be performed on one transformer of each rating.
      4.1.1 Excitation loss
      4.1.2 Copper loss
      4.1.3 Temperature rise per ANSI C57.18
      4.1.4 Double induced voltage at 400 Hz

4.2 Unit tests. Each transformer shall be given the following tests prior to shipment:
    4.2.1 Dielectric test:
        Primary to secondary and ground — 2500 VAC for 10 seconds
        Secondary to primary and ground — 1000 VAC for 10 seconds
    4.2.2 Ratio test

5.0 Data Requirements:
    5.1 Price & Delivery — each option.
    5.2 Estimated size and weight — each option.
    5.3 Sketch of configuration showing location of terminals and mounting method.
    5.4 Estimated no load and loaded losses.
    5.5 Test reports for all type and unit tests must be provided with each shipment.

Technical Standard Document No. TS-35

Standard Title: Storage and Preservation of Skidded Equipment

| Revisions | Date | Prepared By | Approved By | Purpose |

1.0 Scope:
   1.1 This standard is designed to cover the basic steps in preparation of a skid for extended storage. Different climatic conditions may require actions other than those listed below. It is the customer's responsibility to advise Exceltec if the equipment will experience unusual storage conditions.
2.0 Requirements:
   2.1 Exceltec skidded units are designed to operate in severe marine service environments, but special care must be taken to ensure adequate unit protection while in storage prior to installation. Exceltec recommends the following guidelines to ensure that units are protected while in storage.
      2.1.1 To prevent moisture build-up in the control panels, desiccant packs should be added and the panel should be sealed with a heavy plastic cover. If the unit is stored for an extended period of time, more than 2 months, the desiccant should be changed on a 2-month cycle.
   2.2 Rotating Gear:
      2.2.1 All rotating gear should be rotated by hand 10 turns every 2 months, more frequently if possible. This will prevent bearing damage due to storage.
      2.2.2 Use a megger periodically to ensure the integrity of all motor wiring (windings) is maintained. Record all megger readings and investigate any significant drops in insulation resistance.
      2.2.3 The motor bearings are packed with grease from the motor factory. Do not add grease.
   2.3 Store process unit away from excessive heat and corrosives.
   2.4 Store process unit in area away from high risk impacts or jolts.
   2.5 Process unit should be securely tarped to prevent direct sunlight exposure and dirt and debris build-up on unit.
   2.6 Care should be taken not to damage any of the process unit's protective coating, so as to avoid the risk of corrosion.

Index

A

Accident prevention, 6
Acid cleaning
    in electrolysis system, 37
    of electrolyzer cells, 30-31
    of MHI, 33
    of Sanilec system, 33
    in trouble analysis, 96
Acrylonitrile butadiene styrene (ABS), for electrolysis systems, 41, 67
Air cooling, sealed circulating forced, 46. *See also* Cooling methods
Air dilution requirements, 68
Air operated differential pressure systems, for tank level control, 59
Alarms
    D/P device for, 59
    in electrolysis systems, 38
    flow, 39
    in liquid flow control, 52
    temperature related, 58
American Society for Testing of Materials, 75
Ammonia, demand reactions with, 19
Antifungal lacquer, use of, 95
Antimicrobials division, of OPP, 6

B

Basket strainers, 54
Bellows pumps, 53
Beverage operations, disinfection in, 71
Biofouling control, 82
Bipolar electrodes, 29
Bleach
    first use of, 2
    on-site generation of, 2
    powder, 2
Bleaching, industrial, 72
Blower duct arrangements, 68
Blowers, in seawater systems, 27
BPO-DMA (benzoyl peroxide-dimethylaniline) cure, 64
Breakpoint chlorination curve, 18, 19
Brine, salt content testing of, 98
Brine hypochlorite systems, design of, 76-77
Brine system data sheet, 103
Brine electrolyzer systems
    cooling system applications in, 70-71
    fault analysis for, 97
    flow control in, 51, 53
    general commissioning procedure for, 91-92
    general description, 25-26
    horizontal cell circuits in, 36-37
    manufacturers of, 29
    operating costs for, 84-86
    pre-commissioning checklist for, 89-91
    product dosing for, 65-66
    pumps used in, 38
    standard for salt usage in, 85
    start up and operation of, 87-88
    storage tanks, 68
    water softening in, 55-57
Buna-*N*, in electrolysis systems, 42-43
Burns
    eye, 101
    skin, 102

C

Calcarious deposits, in trouble analysis, 96, 98
Calcium carbonate
    in electrolysis system, 37
    scale formed by, 22

Calcium hypochlorite, solid, 16
Calcium waste products, produced by seawater electrolysis, 80
Candidate contaminant list (CCL), USEPA, 12
Candle filters, 54
Cathode cleaning, in seawater electrolyzer maintenance program, 30-31
Cathode deposits, in chlorMaster system, 32
Cell assemblies, in trouble analysis, 96, 97
Cell cleaning, 95
   in brine systems, 26
   in electrolysis systems, 37
   with Sanelec system, 32
   in seawater systems, 27
Chemical price index, 85
Chick-Watson law, 20
Chlorate, formation of, 12
Chlorate ion concentrations, in sodium hypochlorite, 11
Chloride, formation of, 17
Chlorinated polyvinyl chloride (CPVC)
   for electrolysis systems, 39
   limitations of, 67
Chlorination, 19
Chlorine
   commercial production of, 1
   demand reactions for, 17-19
   dumping of, 2
   elemental, 16
   in sodium hypochlorite, 11
   total price of produced, 85, 86
   in water treatment processes, 15-16
Chlorine demand, for seawater hypochlorite systems, 80-81
Chlorine dioxide, disinfectant applications of, 13
Chlorine gas
   discovery of, 1
   disinfection applications for, 9
   production of, 2
   RED on, 6-7
   regulation of, 7-8
Chlorine industry, dimensionally stable anode in, 2
Chlorine liquid
   commercial manufacturing of, 1
   pressurized, 16
Chlorine residual analyzer, 82
Chlorite, regulation of, 12
ChlorMaster electrolytic cell, 31-32
Chloromat electrolyte cells, 30
Chloromat tubular cell, 73
Chlorophenols, formation of, 19
Clams, in electrolysis systems, 78

Clean Air Act (CAA), risk management plan of, 5, 9
Coatings, on steel pump bases, 64
Code review, 78
Concrete dissolver tanks, 60
Control panel
   of electrolysis systems, 49-50
   in seawater designs, 78
Cooling methods
   oil immersion, 47
   for rectifier units, 46
   in seawater designs, 78
Cooling system applications, in seawater systems, 73
Cooling tower applications, disinfectant chemicals in, 15
Cooling water systems, brine system applications, 70-71
Corrosion, consideration of, 75. *See also* Acid cleaning; Cell cleaning
Cost analysis, for seawater system, 86
Coupling guards, 64
CPVC (chlorinated polyvinyl chloride)
   in electrolysis systems, 39
   limitations of, 67
CT values, in disinfection, 15
Cyanide destruction, 72

D

Daiki Engineering, 34
Data sheet
   brine system, 103
   seawater system, 104
DC cell, voltage problem with, 95
DC power
   for brine hypochlorite systems, 76
   equipment for, 46
   in seawater designs, 78
DC rectifier, in electrolysis system, 37. *See also* Rectifier units
Demand reactions
   with ammonia, 18
   chlorine for, 17-19
   with organic and inorganic matter, 19
DeNora, 32, 33, 36, 73
Diamond Shamrock Corp., 2, 32, 73
Diaphragm pump, 53
Dichloramine ($NHCl_2$), 18
Differential pressure (DP) units
   cost effectiveness of, 60
   in electrolysis systems, 50

Index

Dilution fans, on pre-commissioning checklist, 90. *See also* Hydrogen dilution fans
Dimensionally stable anodes (DSA), 28
Disinfection
 kinetics of, 19-20
 principles of, 15
Disinfection applications
 chlorine dioxide, 13
 chlorine gas, 9
 ozone, 13
 sodium hypochlorite, 10-12
Disinfection chemistry
 chlorine application, 15-20
 sodium hypochlorite application, 20-22
Dissolver level controls, 60-61
Dissolver systems, 61
Dosing method, for seawater hypochlorite systems, 81
Dosing point pressure, 79-80
Dosing systems
 formats for, 77
 pump configurations for, 80
D/P switches, 59
Drinking water. *See also* Potable water; Water treatment systems
 incorrect chlorine dose in distribution system for, 11
 treating of, 7
 USEPA CCL for, 12
DSA coating, 32, 33

E

Elastomeric sealing materials, in electrolysis systems, 42
Electrical codes, 78
Electrical control panels, for strainers, 55
Electrical safety, 101
Electrical shock, 102
Electrocatalytic (ELCAT) Corp., 30-31, 35
Electrochemical process, components of, 20
Electrochlorination systems
 installation of, 83
 for mammal pools, 73
Electrochlorinators
 seawater, 27
 strainer arrangements in, 53
Electrolysis, fundamental laws of, 1
Electrolysis system design, 75
 for brine hypochlorite systems, 76-77
 seawater hypochlorite systems, 77-82
Electrolysis systems
 brine system applications, 70-72

brine system water softening, 55-57
 cell level and temperature in, 52-53, 58
 control panel, 49-50
 DC power rectifiers, 44-49
 equipment for, 37-38
 hydrogen handling practices for, 68-70
 inlet seawater strainers, 53-55
 installation of, 83-84
 instruments in, 38-39
 liquid flow equipment, 51-52
 materials of construction for, 39-44
 pipe, valves, and fittings for, 67-68
 pressure and differential pressure equipment for, 50-51
 product storage in, 62-63
 pump equipment for, 63-66
 salt storage-dissolver tanks, 60-62
 seawater system applications, 73
 tank level equipment in, 58-60
 temperature sensing equipment, 57
 water and brine instrumentation for, 53
Electrolytic cell modules
 horizontal cell circuits in, 35-37
 vertical cell circuits, 34-35
Electrolytic cells, 28
 ChlorMaster, 31-32
 Chloromat, 30-31
 components of, 20
 designs for, 29
 hychlorinator, 34
 manufacturers of, 29
 marine growth preventing system of, 33
 plate, 29-34
 reactions, 21-22
 Sanilec, 32-33
 SEACELL, 33
 tubular, 29, 30-31, 34
Electrolyzer cells. *See also* Cell cleaning
 in electrolysis system, 37
 level sensors of, 52
 seawater, 86
 in seawater systems, 27
 temperature sensor of, 52-53
 water supply hardware feeding, 53
Electrolyzers, cost of added, 81
Electrolyzer systems
 brine systems, 25-26
 seawater systems, 27-28
 sensors in, 52
 types of, 25
Elektron Co., 1
Emergency response practices, 6
Engelhard Minerals and Chemicals, Ltd., 30, 73

Environmental Protection Agency, U.S. (USEPA), 75
　office of pesticide programs of, 6-7
　risk management plan of, 5-6
Epoxy resin cement joints, 42
Ethylene propylene diene methylene (EPDM), in electrolysis systems, 42
Exchangers, in brine systems, 26
Eye burns, 101

F

Fasteners, for electrolytic cell modules, 35
Fault analysis
　for brine systems, 97
　electrochlorinator, 95
　for seawater systems, 96
Fiberglass reinforced plastic (FRP)
　dissolver tanks from, 60
　in electrolysis systems, 42, 67-68
Fiberglass reinforced plastic (FRP) tanks, 62, 79
First aid, 101
Fittings, design of, 79
Floating production storage operation (FPSO), 46
Flow control devices, 51
Flowmeters, in electrolysis systems, 39
Flow monitors, 53
Flow sensing, 51, 52
Fluorocarbon elastomer (Viton), in electrolysis systems, 42
Food processing, chlorine disinfection in, 71
Frame assemblies, for electrolytic cell modules, 35

G

Gauge seals, in electrolysis systems, 50
Gear reducers, 55
Gordon-Adam model, 12-13

H

Hall Effect flow sensors, 51, 52
Hastelloy C, 33, 43
Hazardous material, definition for, 8
Heat exchanger, in brine systems, 26
Henry's Law Constant, 16
High density polyethylene (HDPE) tanks, 62, 63

Hom model, 20
Horizontal cell circuits
　in brine electrolysis, 36-37
　in seawater electrolysis, 35-36
Humidity, in seawater designs, 78-79
Hychlorinator electrolytic cell, 34
Hydrochloric acid (HCl), handling of, 99, 100-101
Hydrocyclone, in seawater systems, 69
Hydrogen
　dilution of, 70
　explosive range of, 68
　separated, 69-70
　solubility in hypochlorite product, 68
Hydrogen degassing cyclone, in seawater systems, 27
Hydrogen dilution fans, in electrolysis systems, 38
Hydrogen dilution fan switches, in electrolysis systems, 39
Hydrogen gas seal pot, in seawater systems, 27
Hypochlorite generation system, on-site
　equipment testing requirements for, 88
　installation requirements for, 87, 88
Hypochlorite ion (OCl⁻)
　effect of pH on, 17
　in water, 16
Hypochlorite product, hydrogen solubility in, 68
Hypochlorous (HOCl) acid,
　dissociation of, 16
　effect of pH on, 17

I

Indonesia, labor personnel from, 83
Inflation, chemical price index, 85, 86
Ingestion, as safety issue, 102
Inlet seawater strainers, 53-55
Installation, system, 83-84
Instrumentation
　control panel, 49-50
　on pre-commissioning checklist, 89

J

Joining methods
　with FRP, 42
　with Kynar or Teflon, 41
　with PVC and CPVC materials, 40

Index

K

Kynar (polyvinylidene fluoride), in electrolysis systems, 41, 67

L

Labor, costs of, 85
Level controls
    air operated differential pressure systems for, 59
    dissolver, 60-61
    on pre-commissioning checklist, 90-91
    tank, 58-60
Level sensors, failure of, 52-53, 58
Liquid flow control, 51-52
Local regulations, of hazardous materials, 8
Lower explosive limit (LEL), 68

M

Magnesium, in electrolysis systems, 37
Magnesium hydroxide, produced by seawater electrolysis, 80
Mammal pools, seawater electrochlorination for, 73
Mathieson Chemical Co., 1, 2
M&E index, 85
Metals, in electrolysis systems, 43
Mitsubishi Heavy Industries (MHI), 32, 33
Monitoring, of electrolysis systems, 38
Monochloramine (NH_2Cl), formation of, 18, 19
Monopolar cell, 28, 32
Motors, TEFC (totally enclosed fan cooled), 64

N

National Electrical Manufacturers Association (NEMA)
    classifications of, 46
    electrical code requirements of, 75
Neoprene, in electrolysis systems, 43
Niagara River, 2
Nitrogen trichloride, 18
Noryl plastic material, 56

O

Occupational Safety and Health Administration (OSHA), process safety management standard of, 7-8
Odor control, sodium hypochlorite in, 72
Office of pesticide programs (OPP), of USEPA, 7-8
Offshore systems
    area classification for, 78
    DC power equipment for, 46
Oil field water, 73
Oil immersion cooling, 47
Oil platforms, disinfection applications of, 73
Oronzio DeNora Spa, 32, 33, 36, 73
Oxidation potential (ORP) analysis, in electrolysis systems, 39
Oxidation reduction potential (ORP), 72
Oysters, in electrolysis systems, 78
Ozone, disinfection applications of, 13

P

Pacific Engineering and Construction (Pepcon), 31
Paddle type sensors, 51
Pakistan, labor personnel from, 83
Parasitic reactions, in hypochlorite cell, 22
Pesticide, chlorine gas as, 5
Philippines, labor personnel from, 83
Piping
    design of, 79
    internal distribution and brine removal for, 61
    on pre-commissioning checklist, 89
Pit dissolvers, 60
Pit storage, 63
Plankton blooms, and system stability, 82
Plastic materials. *See also* Fiber reinforced plastic
    in electrolysis systems, 39
    for regeneration valves, 56
Plate cell testing, 32
Plate orifice systems, in electrolysis systems, 50
Plunger designs, 55
Polyehylene (PE), dissolver tanks from, 60
Polypropylene (PP), in electrolysis systems, 40-41, 67
Polytetrafluoroethylene (Teflon), in electrolysis systems, 41, 67
Polyvinyl chloride (PVC), for electrolysis systems, 39-40, 67
Polyvinylidene fluoride (PVDF, Kynar), in electrolysis methods, 41
Potable water, hypochlorite dosing for, 77. *See also* Water treatment systems
Power stations, circulating cooling at, 73

Power supply
  for brine systems, 84
  on pre-commissioning checklist, 89
  three-phase, 80
Pressure control systems, in electrolysis systems, 50
Pressure gauges, in electrolysis systems, 50
Pressure relief equipment, in electrolysis systems, 51
Preventive maintenance program, 95
Process safety management standard, OSHA's, 7-8
Programmable logic control (PLC) system, in electrolysis system design, 77
Proximity switch systems, 59
Pump controls, 66
Pump dosing systems, 65
Pumps
  bellows, 53
  centrifugal, 66
  diaphragm, 53, 77
  in electrolysis systems, 38
  FRP, 64
  magnetically coupled, 63, 65
  positive displacement, 65-66, 77
  seawater booster, 63-65
  seawater hypochlorite dosing, 63-65
  in seawater systems, 27
Pump testing, on pre-commissioning checklist, 89
PVC (polyvinyl chloride), in electrolysis systems, 39-40, 67

R

Rectifier units, 95
  area classification for, 47
  cooling methods for, 45
  independent alarm conditions, 48
  maintenance of, 47
  metering, control, and operation of, 45
  operating status, 48
  safety equipment for, 48-49
Regulation review, 78
Regulations, local and state, 8
Reregistration eligibility decision (RED), on chlorine gas, 6, 9
Resins, in water softeners, 56
Rubber, synthetic, in electrolysis systems, 43

S

Safety, 102
  chemical, 99-101
  electrical, 101
  eye burns, 101
  first aid, 101
  ingestion or gassing, 102
  skin burns, 102
Salt, categories of, 85
Salt content
  and equipment capacity, 81
  testing brine for, 98
  testing seawater for, 95-98
Salt storage-dissolver tanks, 60-62
  dissolver level controls for, 60-61
  internal distribution and brine removal piping in, 61
  salt addition for, 61-62
Sanilec cell modules, 35
Sanilec electrolytic cell, 32
Saudi Arabia, seawater EC sites in, 83
Scale formation, reactions for, 22
Scraper designs, 54
SEACELL electrolytic cell, 33
Seal flush systems, on pre-commissioning checklist, 89
Seawater, salt content testing of, 95-98
Seawater electrolysis
  horizontal cell circuits in, 35-36
  vertical cell circuits in, 34-35
Seawater electrolysis installations, 77-82
Seawater hypochlorite systems
  available water pressure in, 79
  biofouling control in, 82
  dosage control in, 80
  dosing point pressure requirements of, 79-80
  local environmental conditions for, 78
  local regulations and codes for, 78
  power supply requirements of, 80
  site type, 78
  system sizing, 80-82
Seawater system data sheet, 104
Seawater systems
  fault analysis for, 96
  flow control in, 51
  general commissioning procedure, 92-93
  general description, 27-28
  hydrocyclone system used in, 69
  manufacturers of, 29
  operating costs of, 86
  pumps used in, 38
  strainer assemblies for, 53-55
  temperature switches for, 57
Section 112(r), of CAA amendments, 5-6
Separator media, 54
Severn Trent Services, 37
Shock, electrical, 102

Index 205

Skin burns, 102
Snails, in electrolysis systems, 78
Sodium hypochlorite generators, hydrogen handling in, 68
Sodium hypochlorite (NaOCl)
  commercially produced, 10
  decomposition of, 12-13
  handling of, 99-100
  liquid, 16
  on-site generation chemistry for, 20-22
  on-site generation of, 2
  reductions in concentration of, 11
Sodium hypochlorite production, operating economics of, 84
Sodium hypochlorite solution, stability of, 10
Solvent cementing, in electrolysis systems, 40
Sri Lanka, labor personnel from, 83
Stainless steel (316L), in electrolysis systems, 44
State regulations, of hazardous materials, 8
Status indicators, in electrolysis systems, 38
Steel fabrications, 79
Storage tanks, 62-63. *See also* Tanks
  with attached air blowers, 69
  on pre-commissioning checklist, 90
  in seawater systems, 27
Strainers
  basket, 54
  in electrolysis systems, 50
  inlet seawater, 53-55
  in seawater systems, 27
Strainer testing, on pre-commissioning checklist, 90
Supplies, operating, 85
Swimming pools, on-site generators for, 72
Switching power supplies, 44, 45
System design, and trouble analysis, 95
System operations, operating economics of, 480
System sizing
  and equipment requirements, 76
  for seawater hypochlorite systems, 80

T

Taiwan, seawater EC sites in, 83
Tank level control, 58-60
Tank level instruments, in electrolysis systems, 39
Tanks
  FRP, 62, 79
  open top, 68
  redundant, 77
  salt storage-dissolver, 60-62
  storage, 62-63
  water softener, 56
Tap switch voltage control rectifiers, 44, 45
TEFC motors, 64
Teflon (polytetrafluoroethylene), in electrolysis systems, 41, 67
Teflon tape, in electrolysis systems, 40
Temperature
  in brine hypochlorite systems, 76
  and equipment capacity, 81
  in seawater designs, 78
  in trouble analysis, 95, 98
Temperature sensing, equipment for, 57-58
Threaded connections, in electrolysis systems, 40
Thyristor rectifiers, 44, 45
Titanium (Ti), in electrolysis systems, 43
Training requirements, for water and wastewater operators, 7
Trichloramine (NCl_3), 18
Trihalomethanes (THMs), in gaseous chlorine application, 9
Trouble analysis
  for brine systems, 97
  and prevention maintenance program, 95
  for seawater systems, 96

U

Ultrasonic level systems, 59
Ultraviolet (UV) inhibitors
  in FRP, 42
  in PVC and CPVC materials, 40
Ultraviolet (UV) light, disinfection applications for, 9
USFilter Corp., 30

V

Valves, design of, 79
Vane type sensors, 51, 52
Vertical cell circuits, 34-35
Viton (fluorocarbon elastomer), in electrolysis systems, 42
Voltage, cell, in trouble analysis, 97

W

Wastewater operators, training for, 7
Wastewater treatment systems
  chlorine discharged from, 11
  chlorine in, 15-16
  disinfection applications for, 9, 15

hypochlorite dosing for, 77
Water discharge, limitations on, 78
Water disinfection, in brine systems, 71
Water hardness, 56
Water pressure
  available, 79
  in brine hypochlorite systems, 76
Water softener controls, on pre-
      commissioning checklist, 90
Water softeners
  in brine systems, 26, 55-57
  in electrolysis system, 37
  home, 55
  importance of, 57
  resin properties in, 55-56
Water to brine ratio, in electrolysis systems, 39
Water treatment systems
  chlorine in, 15-16
  disinfection in, 15
Wedge wire separators, 54